CONTENTS

HOW EFFECTIVELY IS THE FEDERAL GOVERNMENT ASSISTING STATE AND LOCAL GOVERNMENTS IN PREPARING FOR A BIOLOGICAL, CHEMICAL OR NUCLEAR ATTACK?

FRIDAY, AUGUST 23, 2002

HOUSE OF REPRESENTATIVES,
SUBCOMMITTEE ON GOVERNMENT EFFICIENCY, FINANCIAL MANAGEMENT AND INTERGOVERNMENTAL RELATIONS,
COMMITTEE ON GOVERNMENT REFORM,
Denver, CO.

The subcommittee met, pursuant to notice, at 10 a.m., at the Jefferson County Municipal Building, 100 Jefferson County Parkway, Hon. Stephen Horn (chairman of the subcommittee) presiding.

Present: Representatives Horn and Tancredo.

Staff present: J. Russell George, staff director, chief counsel; Dave Bartel, chief of staff; Bonnie Heald, deputy staff director; Chris Barkley, assistant to the subcommittee; and Michael Sazonov, staff assistant.

Mr. HORN. A quorum being present, this hearing of the Subcommittee on Government Efficiency, Financial Management and Intergovernmental Relations will come to order.

On September 11, 2001, the world witnessed the most devastating attacks ever committed on U.S. soil. Despite the damage and enormous loss of life, the attacks failed to cripple this Nation. To the contrary, Americans have never been more united in their fundamental belief in freedom and their willingness to protect that freedom. The diabolical nature of those attacks and then the deadly release of anthrax sent a loud and clear message to all Americans: We must be prepared for the unexpected. We must have the mechanisms in place to protect this Nation and its people from further attempts to cause massive destruction.

The aftermath of September 11th clearly demonstrated the need for adequate communications systems and rapid deployment of well-trained emergency personnel. Yet despite billions of dollars in spending on Federal emergency programs, there remain serious doubts as to whether the Nation is equipped to handle a massive chemical, biological or nuclear attack.

Today, the subcommittee will examine how effectively Federal, State, and local agencies are working together to prepare for such emergencies. We want those who live in the great State of Colorado and the good people of cities such as Golden and Denver to know that they can rely on these systems should the need arise.

We are fortunate to have witnesses today whose valuable experience and insight will help the subcommittee better understand the needs of those on the front lines. We want to hear about their capabilities and their challenges. And we want to know what the Federal Government can do to help. We welcome all of our witnesses and we look forward to their testimony.

We are very pleased to have with us today as a member of this subcommittee, and without objection, he will have full rights to question and also to have his rights, and that's your own representative, Tom Tancredo. We would like to have an opening statement from him.

[The prepared statement of Hon. Stephen Horn follows:]

HOW EFFECTIVELY IS THE FEDERAL GOVERNMENT ASSISTING STATE AND LOCAL GOVERNMENTS IN PREPARING FOR A BIOLOGICAL, CHEMICAL OR NUCLEAR ATTACK?

HEARING

BEFORE THE

SUBCOMMITTEE ON GOVERNMENT EFFICIENCY, FINANCIAL MANAGEMENT AND INTERGOVERNMENTAL RELATIONS

OF THE

COMMITTEE ON GOVERNMENT REFORM

HOUSE OF REPRESENTATIVES

ONE HUNDRED SEVENTH CONGRESS

SECOND SESSION

AUGUST 23, 2002

Serial No. 107–226

Printed for the use of the Committee on Government Reform

Available via the World Wide Web: http://www.gpo.gov/congress/house
http://www.house.gov/reform

U.S. GOVERNMENT PRINTING OFFICE

87–890 PDF WASHINGTON : 2003

For sale by the Superintendent of Documents, U.S. Government Printing Office
Internet: bookstore.gpo.gov Phone: toll free (866) 512–1800; DC area (202) 512–1800
Fax: (202) 512–2250 Mail: Stop SSOP, Washington, DC 20402–0001

COMMITTEE ON GOVERNMENT REFORM

DAN BURTON, Indiana, *Chairman*

BENJAMIN A. GILMAN, New York
CONSTANCE A. MORELLA, Maryland
CHRISTOPHER SHAYS, Connecticut
ILEANA ROS-LEHTINEN, Florida
JOHN M. McHUGH, New York
STEPHEN HORN, California
JOHN L. MICA, Florida
THOMAS M. DAVIS, Virginia
MARK E. SOUDER, Indiana
STEVEN C. LATOURETTE, Ohio
BOB BARR, Georgia
DAN MILLER, Florida
DOUG OSE, California
RON LEWIS, Kentucky
JO ANN DAVIS, Virginia
TODD RUSSELL PLATTS, Pennsylvania
DAVE WELDON, Florida
CHRIS CANNON, Utah
ADAM H. PUTNAM, Florida
C.L. "BUTCH" OTTER, Idaho
EDWARD L. SCHROCK, Virginia
JOHN J. DUNCAN, Jr., Tennessee
JOHN SULLIVAN, Oklahoma

HENRY A. WAXMAN, California
TOM LANTOS, California
MAJOR R. OWENS, New York
EDOLPHUS TOWNS, New York
PAUL E. KANJORSKI, Pennsylvania
PATSY T. MINK, Hawaii
CAROLYN B. MALONEY, New York
ELEANOR HOLMES NORTON, Washington, DC
ELIJAH E. CUMMINGS, Maryland
DENNIS J. KUCINICH, Ohio
ROD R. BLAGOJEVICH, Illinois
DANNY K. DAVIS, Illinois
JOHN F. TIERNEY, Massachusetts
JIM TURNER, Texas
THOMAS H. ALLEN, Maine
JANICE D. SCHAKOWSKY, Illinois
WM. LACY CLAY, Missouri
DIANE E. WATSON, California
STEPHEN F. LYNCH, Massachusetts

BERNARD SANDERS, Vermont (Independent)

KEVIN BINGER, *Staff Director*
DANIEL R. MOLL, *Deputy Staff Director*
JAMES C. WILSON, *Chief Counsel*
ROBERT A. BRIGGS, *Chief Clerk*
PHIL SCHILIRO, *Minority Staff Director*

SUBCOMMITTEE ON GOVERNMENT EFFICIENCY, FINANCIAL MANAGEMENT AND INTERGOVERNMENTAL RELATIONS

STEPHEN HORN, California, *Chairman*

RON LEWIS, Kentucky
DOUG OSE, California
ADAM H. PUTNAM, Florida
JOHN SULLIVAN, Oklahoma

JANICE D. SCHAKOWSKY, Illinois
MAJOR R. OWENS, New York
PAUL E. KANJORSKI, Pennsylvania
CAROLYN B. MALONEY, New York

EX OFFICIO

DAN BURTON, Indiana

HENRY A. WAXMAN, California

J. RUSSELL GEORGE, *Staff Director and Chief Counsel*
BONNIE HEALD, *Deputy Staff Director*
CHRIS BARKLEY, *Assistant*

ONE HUNDRED SEVENTH CONGRESS

Congress of the United States

House of Representatives

COMMITTEE ON GOVERNMENT REFORM

2157 RAYBURN HOUSE OFFICE BUILDING

WASHINGTON, DC 20515–6143

MAJORITY (202) 225–5074
FACSIMILE (202) 225–3974
MINORITY (202) 225–5051
TTY (202) 225–6852

www.house.gov/reform

Opening Statement
Chairman Stephen Horn
Subcommittee on Government Efficiency,
Financial Management and Intergovernmental Relations
August 23, 2002
Golden, Colorado

A quorum being present, this hearing of the Subcommittee on Government Efficiency, Financial Management and Intergovernmental Relations will come to order.

On September 11th, 2001, the world witnessed the most devastating attacks ever committed on United States soil. Despite the damage and enormous loss of life, the attacks failed to cripple this Nation. To the contrary, Americans have never been more united in their fundamental belief in freedom and their willingness to protect that freedom. The diabolical nature of those attacks and then the deadly release of anthrax sent a loud and clear message to all Americans: We must be prepared for the unexpected. We must have the mechanisms in place to protect this Nation and its people from further attempts to cause massive destruction.

The aftermath of September 11th clearly demonstrated the need for adequate communications systems and rapid deployment of well-trained emergency personnel. Yet despite billions of dollars in spending on Federal emergency programs, there remain serious doubts as to whether the Nation is equipped to handle a massive chemical, biological or nuclear attack.

Today, the subcommittee will examine how effectively Federal, State and local agencies are working together to prepare for such emergencies. We want those who live in the great State of Colorado and the good people of cities such as Golden and Denver to know that they can rely on these systems, should the need arise.

We are fortunate to have witnesses today whose valuable experience and insight will help the subcommittee better understand the needs of those on the front lines. We want to hear about their capabilities and their challenges. And we want to know what the Federal Government can do to help. We welcome all of our witnesses and look forward to their testimony.

EXHIBIT
1
HUNTER & GEIST

Mr. TANCREDO. Thank you, Mr. Chairman, and thank you very much, for—and thanks to the committee for coming to Colorado to discuss these issues. I am pleased because of what I have seen up to this point in time in terms of the way that the State has responded. I am very interested in knowing in more detail exactly how that coordination and cooperation has taken shape in Colorado, as are you and is the Nation.

I hope that the Federal Government is going to be able to provide a certain model for this kind of cooperation through the creation of the homeland defense agency where we, in fact, are attempting to take those agencies that have a responsibility for homeland defense and not just have them cooperating on an interagency basis, but having them part of a single agency with a single purpose, one director, so that everybody seems to be, hopefully, on the same page.

This is, of course—The proposal has passed the house, and I am certainly looking forward to it passing in the Senate. I know that the President has proposed it and is enthusiastically supporting it. So I think it is, in a way, a good model, at least in terms of the way he describes it, the way people can work together. We are not looking for a similar legislatively directed creation of cooperation here in the State, but I think that we can look at what has happened here and, hopefully, around the country and take some hope away from this.

We are going to be asking people here who think about the unthinkable, to help us through this process, and give everyone, I suppose, a feeling of security; not a false sense of security, but a real sense of security because they know that good people put their minds together to come up with programs that will work and be effective. And so I certainly look forward to the testimony today.

And I again want to express my sincere appreciation for you and the committee to come out here and prepare for this.

[The prepared statements of Hon. Tom Tancredo and Hon. Mark Udall follow:]

Committee on Government Reform
Subcommittee on
Government Efficiency, Financial Management,
And intergovernmental relations
Field Hearing
Golden, Colorado
August 23, 2002

Opening Statement for:
CONGRESSMAN TOM TANCREDO

Mr. Chairman, I would like to thank the committee for coming to Colorado to discuss our homeland security. In these days, when we can't be sure where the next attack is coming from we have to make all efforts to keep our country safe and secure.

We have many challenges and must work with the president and one another to achieve that goal.

What we have to do as a country is focus on the hear and now. We have to listen to the people whose job it is to keep us safe and bring assistance when the unimaginable happens.

With the creation of the Homeland Defense Department, we are taking the first step to bring safety and tranquility back to our country. In the Homeland Security Department we will bring the coordination element that is so desperately needed between all our responders.

I thank you.

EXHIBIT
2
HUNTER & GEIST

Committee on Government Reform
Subcommittee on
Government Efficiency, Financial Management,
And Intergovernmental Relations
Field Hearing
Golden, Colorado
August 23, 2002

Statement of
REPRESENTATIVE MARK UDALL

Mr. Chairman, I want to join my colleague, Mr. Tancredo, in welcoming you to Colorado. I regret that a scheduling conflict made it impossible for me to accept your invitation to take part in today's hearing, but I look forward to reviewing the information that will be presented.

The subject of this hearing is most important and most timely.

Since last September 11th, the country and the Congress have been working to respond to the serious challenge of international terrorism.

We are undertaking the largest reorganization of the government in half a century, and we are engaged on a total rethinking of how we approach security and how to plan for the protection of people, places, and things.

This is a tall order. Homeland security has always been an important responsibility of federal, state and local governments. But in the aftermath of the terrorist attacks, the scope of this responsibility has broadened.

Even before last year's attacks, the challenges before us were outlined in an important report from a commission headed by former Senators Gary Hart of Colorado and Warren Rudman of New Hampshire. Unfortunately, it was only after those attacks that the report received the full measure of attention it deserved.

EXHIBIT
3
HUNTER & GEIST

I have reviewed the commission's report carefully and discussed it with Senator Hart, and I have been impressed with the soundness of the report's recommendations. Still, we have to be careful how we take on this challenge.

In establishing the new department, we must work to shape it into a nimble agency that can be quick to respond to the challenges we face.

And we must remember that a vital part of the overall effort will be to assist the states and local governments to help forestall terrorist attacks and, equally important, to respond to attacks should our efforts at prevention not be totally successful.

After all, as we saw in New York, the dedicated people of our states and local communities – firefighters, police, emergency medical personnel, and others – are the first to respond to such emergencies.

But they should not have to do the job alone. The threat is national in scope and the national government must help.

For the assistance is to be most helpful, it must be tailored to the needs and the capabilities of the states and the local governments and must be delivered in a way that does not create new problems.

So, I am very glad that your Subcommittee is looking into how that can be accomplished. I look forward to learning more about the matters you will be exploring, and to working with you in addressing this important subject when Congress reconvenes next month.

Mr. HORN. Thank you very much. Let me now say how we use witnesses. We are delighted that you've put your heart in providing us advance statements before the actual hearing, and they have been very fine. We've looked at all of them, and our staff is here with us. That will become part of a major report with the House of Representatives.

And so the way we operate is, we have an agenda, you see, we start here with the adjutant general. We will be swearing all the witnesses to affirm the oath on the testimony and that will be amended. But when we call on each, automatically the reporter of debates puts that statement in the hearing, and so you don't have to read it. You can get your own thoughts on it in a summary that gets to the essence of your written document.

So don't feel you have to read 10 or 15 minutes. We like it more in the 5-minute range, and then that's good for you and us. So we thank you all for coming and spending your time.

And we will now have you stand and raise your right hand. And if you have any staff behind you that will also talk in answering questions, please have them take the oath too. And the clerk will note and get the names of those, so keep standing back there.

[Witnesses sworn.]

Mr. HORN. Thank you. And the clerk will note that all of them were in the back there and so forth. Just mark the names.

And we are delighted to open, as we have in other hearings, we have Major General Mason C. Whitney, Adjutant General of the Colorado National Guard, a very important portion of how we deal with preventing and solving the terrorism problem.

So, General, thank you for coming.

STATEMENT OF MAJOR GENERAL MASON C. WHITNEY, ADJUTANT GENERAL, COLORADO NATIONAL GUARD, AND EXECUTIVE DIRECTOR, COLORADO DEPARTMENT OF MILITARY AND VETERANS AFFAIRS

General WHITNEY. Mr. Chairman, thank you very much for this opportunity to talk to your committee about the Colorado National Guard and our participation in homeland security. As you are aware, the Colorado Department of Military and Veterans Affairs is the State organization that I am responsible for, and that consists of the Colorado National Guard, which is about 5,000 soldiers and airmen consisting of the Army National Guard and the Air National Guard, as well as the Civil Air Patrol, which is also an important part of our homeland security mission.

Now, that consists of 2,000 volunteers throughout the State of Colorado in 17 different locations, as well as the Veterans Affairs, which consists of over 400,000 veterans within the State of Colorado.

Basically, we have two missions within the Colorado National Guard and, as well, the Civil Air Patrol identifies with those two missions as well, and they include the State mission of emergency response and the Federal mission for national defense. Now, in preparation for those Federal missions, we receive about $135 million per year for the Colorado National Guard from Federal funds to train and participate in those Federal missions of national defense. The State mission response, we receive about $4.2 million of State funds to prepare for State emergency response missions.

When September 11th arrived, and the terrorist events that took place then, we immediately recalled our Air National Guard units so that they would be prepared to launch F16s and maintain air supremacy or take care of any kinds of activities that still may be or may have been prevalent during the missions that we saw that were accomplished by the terrorists.

So we had, within 15 minutes, two F16s prepared to launch after the terrorist events on September 11th. And within 30 minutes of those terrorist events, we had both F16s airborne, and we had all 16 of our F16s at Buckley Air Force Base prepared in case they were needed.

Along with that, we also had our Army National Guard units recalled that would be providing any kind of security reinforcements for law enforcement agencies throughout the State. All in all, we had over 1500 of our air National Guard and Army National Guard personnel that were at their duty stations within approximately 45 minutes of those terrorist activities.

As you know, we also were called upon to provide airport security in support of the law enforcement organizations that were responsible for the airport security throughout the State of Colorado. We had over 220 of our soldiers that were on active duty, that were on State duty with Federal pay, for over 8 months during that period. They provided airport security to over 13 airports throughout the State of Colorado.

During the time that we have responded to this homeland security mission, we have performed over 4,000 days of State active duty in response to other State emergencies, such as forest fires that have been prevalent in the State of Colorado this year. So, as you can see, it's been a very busy year so far for the Colorado National Guard.

Now, of course, the Civil Air Patrol has capabilities as well to respond to homeland security missions. Visual reconnaissance throughout the State, coupled with their search and rescue capabilities, is one of their main missions that they perform within the State of Colorado.

We have some unique capabilities within the National Guard also that aid in our homeland security mission, and one of those unique missions that we have is a Weapons of Mass Destruction Civil Support Team that consists of 22 full-time, 100 percent federally funded employees of the Colorado Army and Air National Guard that respond to nuclear, biological, and chemical events throughout the State, as well as a geographical area that used to be defined as FEMA Region VIII, but now they've redefined those regions to more accurately—or more adequately deploy other civil support teams throughout the United States so we have 100 percent coverage with every State.

That team and the teams also, the other 26 teams throughout the United States, have been extremely busy over the last several months. They have responded to anthrax precautions that were taken by local agencies or by the FBI. We've also responded to events that could be considered a chemical liability within the State. We've supported other law enforcement and local responders with that team in terms of exercises, making sure that we are compatible with their operational procedures.

All units of the Colorado Army and Air National Guard are also trained in every possibility of warfare, and one of those obviously is chemical warfare, biological warfare, and nuclear warfare. That's a matter of their readiness training, so it's something that comes second nature to them. So there is response capability within the entire organization for those type of events.

Sir, that basically summarizes what we've done over the last several months in terms of homeland security, in terms of what our capabilities are.

We also have some challenges ahead, and one of the challenges we face, I think, is: Is the National Guard properly equipped and trained to be able to respond to those types of terrorist events that could happen in the future? That's something I think that will be on a major agenda item for our new commander of Northern Command, U.S. Northern Command down at Colorado Springs, General Ed Eberhart.

Mr. HORN. You might want to spell that.

General WHITNEY. Yes, sir. E-b-e-r-h-a-r-t.

And General Eberhart, obviously, has been in discussions with the National Guard already about what our role will be with the homeland security mission within the military, realizing that General Eberhart operates in a Title 10 world, the Title 10 Federal statutes, and the National Guard operates in the Title 32 world, which is essentially the State part of the Federal statutes.

Subject to your questions, that's all I have, sir.

[The prepared statement of General Whitney follows:]

Major General Mason C. Whitney, The Adjutant General of Colorado
Executive Director of the Colorado Department of Military and Veterans Affairs

Testimony for the House Committee on Government Reform's Subcommittee on Government Efficiency, Financial Management and Intergovernmental Relations

Date: Friday, August 23, 2002 Place: Jefferson County Municipal Building, Golden, CO

Federal Government Assistance to State Government in Preparing for Biological, Chemical, or Nuclear Agent Attacks

The Federal Government supports the State of Colorado in preparing for terrorist attacks involving biological, chemical, or nuclear agents through organizing, funding, equipping and establishing training standards for Colorado National Guard units capable of responding to these events. National Guard units are uniquely organized as a state controlled military that can be used as an emergency response force by the Governor and can additionally be called to active duty as a reserve force for national defense missions. While still under state control, these units can also be federally funded and equipped to train for their national defense role or can be tasked to perform federal operational missions in their state as requested by the President with the concurrence of the Governor. When they are in this state status, the National Guard is not subject to Posse Comitatus and is able to exercise a level three peace officer authority. They can then be used in support of law enforcement operations throughout the state, regardless of pay status.

Colorado was the first state to field a mission ready National Guard Weapons of Mass Destruction/Civil Support Team (WMD/CST). This team consists of twenty-two full time highly trained soldiers and airmen who are equipped with state of the art equipment used to detect and mitigate nuclear, biological, or chemical (NBC) events. This team works closely with the FBI, state agencies, and local authorities in responding to these NBC events throughout an assigned geographical area. They can also be used in a federal status to respond to NBC events throughout the United States. They are one hundred percent federally funded and are evaluated by the United States Army for mission readiness. They were used extensively during the most recent anthrax threats. There are presently twenty-seven National Guard WMD/CST teams throughout the United States.

All units of the Colorado Army and Air National Guard are trained to operate in a NBC environment as a measure of their readiness for their national defense mission. Although their federal equipment is dated and has not kept up with the emerging NBC threats, they do have a limited capability to respond to NBC events with a large percentage of their force. The federal government provides one hundred percent of the funds for the NBC training and equipment received by the National Guard. Like the WMD/CST unit, these combat units can be used in a state status to respond to state emergencies, such as NBC terrorist attacks. They also have a capability of responding in a state status using federal funds and equipment with the permission of the President and the concurrence of the Governor. They could also be called to active duty and placed in a federal status under federal control.

As we better define the roles our military organizations will play in the evolving threat of terrorism within our borders, we should ponder two significant thoughts. The first thought has to do with the relationship between U.S. Northern Command and the fifty-four National Guard organizations. The National Guard is forward deployed to over 3,000 communities throughout the United States, the District of Columbia, and three territories. It is uniquely capable of supporting local authorities with an immediate emergency response under command and control of the Governor. The National Guard can also be federalized to support USNORTHCOM in their Homeland Security mission, if necessary. When is it necessary for USNOTRHCOM to federalize the National Guard? Under what authority does USNORTHCOM mobilize the National Guard for the Homeland Security mission?

The second thought to ponder concerns the organizing, training, and equipping of the National Guard for the Homeland Security mission. The initiative of the Total Force policy, implemented over thirty years ago, has allowed the Air and Army National Guard to maintain high levels of readiness for their national defense mission. This has been accomplished with varying degrees of success by the U.S. Air Force and the U.S. Army. As a result of this policy, the National Guard has received federal funds and equipment that have allowed it to maintain a relevance to its federal mission. The National Guard has been

EXHIBIT
4
HUNTER & GEIST

integrated into the active duty commitments for global military response because of the downsizing of the active military forces and the high operations tempo that currently exists. The result has been an unprecedented use of the National Guard over the past few years in contingency and peacekeeping operations throughout the world. Some say the National Guard should focus exclusively on the Homeland Security mission at the expense of the global national defense mission because of a concern about over tasking our citizen-soldiers. I think this would be a serious mistake. It would divorce the National Guard from the Total Force and reduce the capability of the United States to respond to global threats with a cost effective reserve force. The National Guard would become a less capable force as a result of the loss of training and equipment that has defined our contribution to our national defense. Any Homeland Security mission should be in addition to the national defense missions already established within the National Guard training programs. New equipment and training for Homeland Security could be integrated into training programs to allow for an adequate response capability within existing units. New units with anti-terrorist capabilities like the National Guard's WMD/CST should be researched and funded if they prove to be an effective counter to terrorist threats.

The National Guard training and equipment funded by the federal government has laid a good foundation for limited responses to terrorist attacks involving nuclear, biological, or chemical agents. USNORTHCOM can build on this capability with research and development focused on Homeland Security. New and existing military organizations that are properly equipped and trained for this evolving mission can support state and local governments in meeting the challenges of protecting our American citizens at home.

Mr. HORN. And we will now go to Ms. Mencer. Ms. Mencer is the executive director, Department of Public Safety, director of Homeland Security, State of Colorado. I assume that is the Governor's sort of operation?

Ms. MENCER. That's correct.

Mr. HORN. Put it all together?

Ms. MENCER. Yes.

STATEMENT OF SUZANNE MENCER, EXECUTIVE DIRECTOR, DEPARTMENT OF PUBLIC SAFETY, AND DIRECTOR, HOMELAND SECURITY, STATE OF COLORADO

Ms. MENCER. Thank you, Mr. Chairman. Thank you for inviting me to speak here today.

I think we realized in 1993, after the first bombing of the World Trade Center, and then after the bombing of the Murrah Building in 1995, that we needed to refocus our efforts on this issue of terrorism, particularly domestic terrorism, which we, I don't think, gave much attention to at all until 1995. I think what we learned on September 11, 2001, is that our efforts still were not sufficient to combat this threat, either internationally or domestically.

I have the unique advantage, I guess, of having been in the FBI for 20 years; 13 of those years I was a counterterrorism and foreign counterintelligence supervisor, both at FBI headquarters and here in Denver. So I have that in my background as well, which serves me well now that I try to look at how to protect the State with not only sharing intelligence with law enforcement agencies, but also with looking at plans, preparedness, and response issues as well.

On November 7, the Governor, by executive order, created the Office of Preparedness and Security in the State. Given our tight fiscal problems that we are having now, he created this office using existing resources that the State already had, and focused our efforts in the area of terrorism. To do this, we used my personnel from the Department of Public Safety, we used—we borrowed one person from the Department of Health and Environment that specialized in bioterrorism, and we used two people from the Office of Emergency Management, who have always done a good job at protecting the State in all kinds of all-hazards approaches to disasters.

But their specific tasking has been to look at focusing the State on preparedness issues of weapons of mass destruction. To do that, we have divided this office into two areas of focus. One is preparedness, plans, and response. To do that—and this office, by the way, was signed into law by the Governor after passing the legislature in this last legislative session in the Colorado legislature, so it became an office permanently in June.

The focus of this office then is twofold. One is for plans, preparedness, and response. To do that, the Governor announced 2 weeks ago that we were dividing the State into seven districts. Six of those districts are the existing State Patrol districts with one exception, and that is, we carved out the five counties of the metro area to create the seventh district. We didn't want the six districts to be overwhelmed by the key assets that we have up and down the front range. So we decided to create this seventh district of the metro area, which we did.

These districts will then be coordinated by three individuals: A major of the State Patrol of that district, an OEM regional planner that is assigned to that area, and a CBI investigator, agent-in-charge. Their responsibility will be to bring to the table everyone from that district that should have a say in plans and preparedness issues. They will look to fire chiefs, to police chiefs, to sheriffs. They will look to health care professionals in that district. They will look to emergency first responders in that district, and ask for assistance from the National Guard and FEMA to coordinate what should the plans be, are the existing plans sufficient. And, indeed, we already have good plans out there; we don't want to reinvent the wheel. Are they sufficient to handle weapons of mass destruction? If not, what additional responses do they need in that area, and what kind of training and equipment do they need that they don't already have? So we'll be looking at that in each of the districts.

Additionally, we'll also be looking at developing an intelligence collection piece. I think when I was in the FBI, sharing of intelligence with local law enforcement agencies consisted of me receiving the information from the sheriffs and chiefs and saying, "Thank you." That was about the end of the sharing. I think what—is there clapping back there?

I think what we've learned is that we have to share better. And, indeed, the FBI created joint terrorism task forces all over the United States. Denver was one of the first offices to do that. I'm sure Mr. Carballido will speak more about that. And I'm pleased to say that we have a Colorado State trooper on the domestic side of the joint terrorism task force. So we are very lucky to have a seat at the table.

Unfortunately, the task forces, by necessity, are limited as to who can be at the task forces. So we need to have a way to better have the chiefs and the sheriffs and other folks out there that have intelligence or that see things every day to be able to share that. So we are going to act as a clearinghouse for this kind of information so that we can then package it and provide it to the FBI if it rises to the level of an investigative concern. So that's what we hope to do.

We do need to break down some barriers among law enforcement agencies. It is difficult, if you work hard to get information, to then give it up and share it with other agencies. That's always been a problem with law enforcement. So we are looking at ways to try to break down those barriers because I think what we've all learned since September 11th is we do need to share better, and we are going to work very hard at doing that.

We applaud the efforts of the Department of Homeland Security and of Governor Ridge, and the President's initiative to create this new department, similar to what we have done here in the State, taking existing resources, combining them together to focus on this issue of terrorism. So we very much applaud that.

We are in close contact with the Office of Homeland Security. I am in biweekly conference calls with them. I just had one yesterday. We share their concerns. They're divided by FEMA regions. We are with FEMA Region VIII and X on our conference calls. And all the States' homeland security directors are there. So it is an opportunity for us to share our concerns, ask our questions, and get answers at that time, and they've been very responsive to all the questions we've had. So we have a good relationship with them.

We also—I go back about every other month to meet with Governor Ridge and his staff. All the homeland security directors do, and this has been very helpful as well, so that we know where they're going, what their direction is, and what kind of information they can provide to us to better make our system here in Colorado more functional in working with them. So we are busy on that.

The Governor also created an infrastructure committee 2 weeks ago, and this will be to bring in the private sectors as well in the State. We will have at least 13 people at that table, each representing the 13 critical infrastructures as defined by the national security strategy. We will look at those 13 infrastructures and then add some as well, such as education, to decide things for the State, like the structure of the threat level system. What does it mean for education if we go up from a Level Yellow, which we are at now, to a Level Orange? What does it mean—do you send your kid to school if we go up to a Level Red? Do you ride the light rail if we are at the red level?

These are the kinds of questions we will ask this committee to come up with by asking their constituents to come up with the models that we need to use and then disseminate to the public as to how we are going to react to this kind of thing. So that way we will bring in the private sector as well and get input from everyone because this is indeed a problem that has to be solved by everyone, not just law enforcement, not just the National Guard, not just Public Safety, but we have to all work together on this. Thank you.

[The prepared statement of Ms. Mencer follows:]

cdps
COLORADO
DEPARTMENT
OF PUBLIC SAFETY

Office of the Executive Director
700 Kipling St.
Suite 1000
Denver, CO 80215-5865
(303) 239-4398
FAX (303) 231-9708

When a truck bomb detonated on February 26, 1993, at 12:18 p.m. in sub-basement B-2 of the World Trade Center, we in the FBI knew that we had to work harder. We had to work harder to develop sources, harder to develop intelligence, harder to gather information. When another truck bomb detonated on April 19, 1995, in front of the Alfred P. Murrah Federal Building in Oklahoma City, we once again vowed to never let it happen again. After both of those, although the resolution was firm in all our minds, the structure of the federal government remained the same.

While we all knew that the ability to thwart another attack, whether foreign or domestic depended on excellent sources, methods and the ability to collect and share intelligence, the mechanism to do that was antiquated.

As an FBI Special Agent for 20 years, and a supervisor of Foreign Counter Intelligence and Counter Terrorism for 13 of those years, I know that the FBI did not consider state and local law enforcement to be equal partners in this effort. Not until the FBI formed the Joint Terrorism Task Forces (JTTF) did local enforcement have an equal seat at the table. Unfortunately, those local law enforcement agencies that are members of the JTTF are, by necessity, few. I'm pleased that the Colorado State Patrol has a Trooper on the Denver Joint Terrorism Task Force.

Now, as the Executive Director of the Department of Public Safety for the State of Colorado, I know that the rest of law enforcement in the state needs to have a mechanism for collecting, disseminating and receiving intelligence from and with the FBI and other federal agencies. The degree to which we shared information before September 11th, was not sufficient. President Bush has requested the creation of the Department of Homeland Security. A new department to combine existing agencies to focus their efforts and their expertise to detect, deter and prevent terrorist acts and respond when necessary. This new department will provide a new focus to collect, collate and disseminate intelligence. It will combine the existing federal resources that have responsibility for protecting our borders, for responding to a disaster, for collecting intelligence for a common goal – keeping us safe from those who would harm us. I applaud the President and Governor Ridge and his staff for their vision and efforts.

Bill Owens
GOVERNOR

C. Suzanne Mencer
EXECUTIVE DIRECTOR

Colorado State Patrol

Colorado Bureau of Investigation

Division of Criminal Justice

Division of Fire Safety

Our Governor, Bill Owens, by executive order on November 7th, created the Office of Preparedness, Security and Fire Safety. The Colorado legislature made it law. This new colorado office has two missions: to plan, prepare and respond as a state to a terrorist act; and to collect and share intelligence among state and local law enforcement agencies and first responders and with the FBI. We have divided the state into seven districts, each coordinated by a State Patrol Major, an OEM regional planner and Colorado Bureau of Investigation Agent in Charge. They will coordinate in each district with chiefs, sheriffs, fire chiefs, health care professionals and emergency first responders. They will review response plans, assess resources and training needs and collect and disseminate

EXHIBIT 5 HUNTER & GEIST

intelligence to prevent and deter future attacks. We need their cooperative effort. Each law enforcement agency, first responder and health care professional needs to work together. We at the state and local level are hoping for even greater cooperation and intelligence sharing with the FBI.

All law enforcement learned on September 11th, that our prior efforts were not enough. We've also learned that our first responders need to have access to more intelligence. They need to be equal partners. We are making great progress, but creating plans, assessing needs, providing training come with a cost. We, as a state, need additional resources to make our state more secure. As with other states, our revenues are down. We can't look to our own coffers to answer our needs. We are anxiously anticipating the federal money from the supplemental to enhance our capabilities. I participate in regular conference calls with Governor Ridge and his staff. They have been very helpful and responsive. We are all working together to ensure that we are safe. We cannot do it alone

C. Suzanne Mencer
Executive Director
Colorado Department of Public Safety

Mr. HORN. Thank you. That's very helpful. And in the question period I'm sure we'll have plenty to check. And we are now having Raul Carballido, who is the acting special agent-in-charge, Federal Bureau of Investigation. Is that in the Denver office?

Mr. CARBALLIDO. Yes, sir.

Mr. HORN. Well, we are delighted you and your other people, starting with Y2K years ago and our jurisdiction over Federal executive computers. And they have been greatly helpful with their hacking and all the rest of it, and on the lootists and the viral this or that and sickness here and there in computers; and they've really been a great help for private industry as well as for government. And so thank you for being here.

STATEMENT OF RAUL E. CARBALLIDO, ACTING SPECIAL AGENT IN CHARGE, FEDERAL BUREAU OF INVESTIGATION

Mr. CARBALLIDO. Thank you, Mr. Chairman. Good morning, Chairman, board members of the subcommittee, and distinguished members of the Colorado delegation. I value the opportunity to appear before you and discuss terrorism preparedness, including threats posed by attacks involving weapons of mass destruction, as well as measures being taken by the FBI and law enforcement partners to address these threats. The mission of the FBI's counterterrorism program is to detect, deter, prevent, and swiftly respond to terrorist actions that threaten U.S. interests at home or abroad, and to coordinate those efforts with local, State, Federal, and foreign entities as appropriate. The Denver field office of the FBI is responsible for the States of Colorado and Wyoming. The field office's headquarters is located in Denver with satellite offices throughout the States of Colorado and Wyoming.

Enhanced cooperation among law enforcement at all levels is a significant component of the prevention and investigation of terrorism. This cooperation is most evident in the development of the Joint Terrorism Task Forces, short JTTF, that now exists in all 56 FBI field offices. These task forces are successful for the integration of resources provided by local, State, and Federal agencies.

The Denver field office, JTTF, was formed in 1996. It is composed of numerous Federal, State, county, and municipal law enforcement agencies. The Denver JTTF also coordinates with the recently created Colorado Office of Preparedness, Security and Fire Safety to share information on terrorism-related matters. This cooperation is demonstrated through the anticipated sharing of an analyst who will conduct terrorism-related research and analysis for the FBI and the Colorado Office of Preparedness, Security and Fire Safety.

Denver, as you know, Mr. Chairman, is also home to a number of significant military assets located in Colorado and Wyoming. The newly established Northern Command, NORTHCOM, is now located at Peterson Air Force Base in Colorado Springs. On October 1, NORTHCOM, Homeland Security Combatant Command, will become operational. In an effort to facilitate cooperation, coordination, and the sharing of information between the FBI and NORTHCOM, in issues relative to terrorism activities and other activities of joint interest, the FBI will assign a full-time, senior-level special agent to the command.

The Denver field office of the FBI has taken a proactive approach in its preparation for terrorist attacks involving weapons of mass destruction. In January 1999, Denver field office completed its own Weapons of Mass Destruction Incident Contingency Plan. This plan was designed to provide specific guidance for response within the Denver field office. The plan is updated annually and has been utilized numerous times to resolve weapons of mass destruction threats or incidents within Colorado and Wyoming.

In addition to the domestic preparedness training that we have received over the years, we have also provided weapons of mass destruction awareness and response training to numerous law enforcement, fire, emergency medical, emergency management, military and infrastructure agencies and organizations.

The Denver field office has also participated in numerous local, regional, and national weapons of mass destruction exercise scenarios. We consider our involvement in both training and exercises to be essential to maintaining and enhancing our relationships with our local, State, and Federal partners.

In 1996 the FBI established a Hazardous Materials Response Unit, which is based in Quantico, Virginia. Staffed with subject matter experts, the unit has provided national and international assistance in the response to weapons of mass destruction terrorism. In May 2000, the unit certified the Denver field office's hazardous materials response team. This is one of 17 teams throughout the country and a regional asset for the FBI. This team, which is comprised of FBI agents specially trained to operate in a contaminated environment, has the capability to respond to a crime scene where weapons of mass destruction may be present.

We have also participated in the development of local weapons of mass destruction response plans. An important example of this is the recent development of a Metropolitan Medical Response Plan for the Denver Metropolitan area and the ongoing development of a similar plan for Colorado Springs.

In December 1998, FBI Denver took a leadership role in the establishment of what is known as the Colorado Counterterrorism Advisory Council. This group, which has met monthly since its inception, includes representatives from a variety of State and Federal agencies who have primary responsibility for response to weapons of mass destruction issues within the State of Colorado. The group has also established interagency notification and response protocols which have greatly enhanced our ability to efficiently share information and provide response resources.

Since the deliberate distribution of anthrax in the U.S. mail during December and October 2001, FBI Denver has responded to more than 800 telephonic requests for assistance concerning potential weapons of mass destruction terrorism, primarily regarding anthrax. In addition, FBI Denver provided field responses to potential weapons of mass destruction incidents on more than 100 occasions and opened more than 30 criminal investigations. The ability to handle this many incidents is a direct result of the partnerships developed among the Federal, State, and local response community, which included the establishment of efficient operational protocols and agreements for laboratory testing of chemical, biological, and radiological agents.

In conclusion, Mr. Chairman, terrorism represents a continuing threat to the United States, and as the lead Federal agency for response to a weapons of mass destruction threat or incident, the FBI must remain prepared to tackle this formidable threat—formidable challenge, I should say. In order to effectively and efficiently respond to the threat, the Denver field office of the FBI continues to enhance its counterterrorism program.

Chairman Horn, this concludes my prepared remarks. I would like to express my appreciation to this subcommittee's examination of the issue of counterterrorism preparedness, and I look forward to responding to any questions. Thank you, sir.

[The prepared statement of Mr. Carballido follows:]

STATEMENT OF RAUL E. CARBALLIDO
ASSISTANT SPECIAL AGENT IN CHARGE, DENVER DIVISION
FEDERAL BUREAU OF INVESTIGATION
BEFORE THE HOUSE COMMITTEE ON GOVERNMENT REFORM
SUBCOMMITTEE ON GOVERNMENT EFFICIENCY,
FINANCIAL MANAGEMENT AND INTERGOVERNMENTAL RELATIONS
AUGUST 23, 2002

Good morning Chairman Horn, Members of the Subcommittee and distinguished Members of the Colorado Delegation. I value the opportunity to appear before you and discuss terrorism preparedness, including threats posed by attacks involving biological, chemical or nuclear agents, as well as measures being taken by the FBI and our law enforcement partners to address these threats.

Introduction

The mission of the FBI's Counterterrorism Program is to detect, deter, prevent, and swiftly respond to terrorist actions that threaten U.S. interests at home or abroad, and to coordinate those efforts with local, state, federal and foreign entities as appropriate. The counterterrorism responsibilities of the FBI include the investigation of domestic and international terrorism. As events during the past several years demonstrate, both domestic and international terrorist organizations represent threats within the borders of the U.S.

The Denver Field Office of the FBI

The Denver Field Office of the FBI is responsible for the States of Colorado and Wyoming. The Field Office has investigative responsibilities in 87 counties with an approximate population of 4.9 million. The Field Offices' headquarters is located in Denver with satellite offices, or Resident Agencies (RAs) in Boulder, Colorado Springs, Durango, Fort Collins, Glenwood Springs, Grand Junction and Pueblo, Colorado, and Casper, Cheyenne, Jackson Hole and Lander, Wyoming. The Field Office personnel resource staffing level for Special Agents is 132 and the professional support complement is 96.

EXHIBIT
6
HUNTER & GEIST

Joint Terrorism Task Forces (JTTFs)

Enhanced cooperation among law enforcement at all levels is a significant component of the prevention and investigation of terrorism. This cooperation is most evident in the development of the Joint Terrorism Task Forces that now exists in all 56 FBI field offices. These Task Forces are successful through the integration of resources provided by local, state and federal agencies.

Additionally, the U. S. Attorney General has directed the U. S. Attorney in each judicial district to form an Anti-Terrorism Task Force (ATTF). This task force coordinates with the JTTF to avoid duplication of effort and enhance the exchange of information and overall counterterrorism objectives.

Denver JTTF

The Denver Field Office JTTF is composed of the Air Force Office of Special Investigations, U.S. Customs Service, U.S. Secret Service, Immigration and Naturalization Service, Drug Enforcement Administration, Colorado State Patrol, U.S. Postal Service, Bureau of Alcohol, Tobacco and Firearms, Bureau of Reclamation, Jefferson County Sheriff's Office, Denver Police Department, Aurora Police Department, Diplomatic Security Service and the Internal Revenue Service. Soon to join the Denver JTTF are deputies from Douglas County and Arapahoe County Sheriff's Offices.

Denver also maintains close liaison with the Defense Intelligence Agency.

In addition to the ATTF, the Denver JTTF coordinates with the recently created Colorado Office of Preparedness, Security and Fire Safety to share information on terrorism-related matters. This cooperation is demonstrated through the anticipated sharing of an ATTF analyst who will conduct terrorism-related research and analysis for the FBI and the Colorado Office of Preparedness, Security and Fire Safety.

Military and Civilian Assets

Denver is also home to a number of significant military assets located in Colorado and Wyoming. Those facilities include Army Space Command, Air Force Space Command NORAD, U.S. Space Command, Buckley Air Force Base, F.E. Warren Air Force Base, Peterson Air Force Base, Schriever Air Force Base, U.S. Air Force Academy and Fort Carson Army Base.

In addition, the newly established Northern Command (NORTHCOM) is now located at Peterson Air Force Base. On October 1, 2002, NORTHCOM, the Homeland Security Combatant Command, will become operational. NORTHCOM will be based in Colorado Springs, Colorado, under the command of United States Air Force General Ralph E. Eberhart. The new command will place all U.S. military air, land, and sea forces and antiterrorist teams charged with protecting the United States under a single entity. In an effort to facilitate cooperation, coordination, and the sharing of information between the FBI and NORTHCOM, in issues relative to the prevention, deterrence and detection of terrorism activities threatening the national defense, and other activities of joint interest to the FBI and NORTHCOM, the FBI will assign a full-time senior level Special Agent to the command. The primary responsibility for those involved in this joint effort is to provide an effective interface between all FBI and NORTHCOM components so as to insure the timely sharing of pertinent law enforcement and counterespionage/counterintelligence information. Through this partnership, the missions of both entities can be completed in a manner that will best serve the people of the United States.

Denver also relies on liaison with many other facilities, such as the National Renewable Energy Lab, Lockheed Martin Technologies, the Plague Center, Rocky Flats, University of Colorado, Colorado School of Mines, Colorado State University, University of Wyoming, the Olympic Training Center, the Pueblo Chemical Storage Depot, numerous ski and mountain resorts and numerous companies engaged in research and development in the Denver Tech Center.

Bioterrorism and Weapons of Mass Destruction

In June 1995, Presidential Decision Directive 39, "the United States Policy on Counter-Terrorism" was issued as a directive for the federal response to any terrorism incident, including those involving chemical, biological, radiological or nuclear weapons, also known as Weapons of Mass Destruction. Presidential Decision Directive 62 reaffirmed the FBI's role as the Lead Federal Agency for conducting and coordinating the federal crisis management response concerning any terrorism incident on U.S. soil, including any incident involving Weapons of Mass Destruction. Crisis management involves measures taken to identify, acquire, and plan for the use of resources needed to anticipate, prevent and/or resolve an act of terrorism.

The Denver Field Office of the FBI has taken a proactive approach in its preparation for terrorist attacks involving Weapons of Mass Destruction.

Realizing that an effective response to any Weapons of Mass Destruction threat or incident would require a multi-jurisdictional, multi-disciplinary effort, the Denver Field Office of the FBI has worked to establish long-term working relationships with our federal, state and local response partners. These efforts have been multi-faceted, including development of a Denver Field Office Weapons of Mass Destruction Incident Contingency Plan, participation in interagency training and exercises, participation in the development of regional response resources and plans, development of interagency working groups, and establishment of operational protocols for effective and efficient response to Weapons of Mass Destruction threats or incidents. Each of these efforts will be discussed in further detail.

Denver Field Office Weapons of Mass Destruction Incident Contingency Plan

In 1998, the Domestic Terrorism Section located at FBIHQ produced a national Weapons of Mass Destruction Incident Contingency Plan. This plan was designed to facilitate the FBI's response to Weapons of Mass Destruction threats or incidents by providing the general framework for evaluating the threat and managing the crisis. In January 1999, the Denver Field Office completed its own Weapons of Mass Destruction Incident Contingency Plan. This plan was designed to provide specific guidance for response within the Denver Field Office. Perhaps the most important aspect of the Denver Field Office plan was that, in addition to summarizing the national response plan, it contained specific contacts for our response partners within Colorado and Wyoming. The contacts identified in this plan, which are updated annually, have been utilized numerous times to resolve Weapons of Mass Destruction threats or incidents within Colorado and Wyoming.

Training and Exercises

Due to the increasing threat that terrorists could utilize Weapons of Mass Destruction within the United States, Congress appropriated funds in 1996 to improve the response to these types of events. Nationally, 150 cities were selected for what was known as the "Domestic Preparedness Program", with Denver being among of the first to benefit from this program. Since 1996, the Denver Field Office of the FBI has participated in three iterations of Domestic Preparedness Program training in Denver, Aurora and Colorado Springs.

In addition to the Domestic Preparedness Training, FBI Denver has provided Weapons of Mass Destruction awareness and response training to numerous law enforcement, fire, emergency medical, emergency management, military and infrastructure agencies and organizations. These presentations have been instrumental in establishing and maintaining meaningful liaison with our federal, state, local and private response partners.

The Denver Field Office has participated in numerous local, regional and national Weapons of Mass Destruction exercise scenarios. The most extensive participation, both in time and resources, was during the national Weapons of Mass Destruction exercise known as TOPOFF 2000. TOPOFF 2000 was a 10 day, no-notice, national exercise which included a scenario where the biological pathogen responsible for plague was deliberately released in the City of Denver. Over the next 10 days the Denver Field Office of the FBI worked with a wide variety of response agencies in our role as lead federal agency for crisis management, and as a result we were able to evaluate and improve our ability to respond to a bioterrorism attack.

We continue to actively participate in Weapons of Mass Destruction training and exercises. During the month of August 2002 alone, we will participate in conferences and exercises sponsored by the Department of Justice, the FBI, the Department of Defense, the Department of Energy, and the City of Denver. We consider our involvement in both training and exercises to be essential to maintaining and enhancing our relationships with our local, state and federal partners.

Response Plans and Resources

By definition the threatened or actual use of Weapons of Mass Destruction involves hazardous chemical, biological, radiological or nuclear materials. Therefore, responding to a threatened or actual release of such materials requires specialized equipment and training.

In 1996 the FBI established a Hazardous Materials Response Unit, which is based in Quantico, Virginia. Staffed with subject matter experts, the Hazardous Materials Response Unit has provided national and international assistance in the response to Weapons of Mass Destruction terrorism. In May 2000, the Hazardous Materials Response Unit certified the Denver Field Office's Hazardous Materials Response Team. This team, which is comprised of FBI Agents specially trained to operate in a contaminated environment, has the capability to respond to a crime scene where Weapons of Mass Destruction may be present.

In addition to developing specialized response resources, the Denver Field Office of the FBI has participated in the development of local Weapons of Mass Destruction response plans. An important example of this is the recent development of a Metropolitan Medical Response Plan for the Denver Metropolitan Area and the ongoing development of a similar plan for Colorado Springs. These plans establish protocols and relationships between emergency medical, hazardous materials, police, fire, hospital, and emergency management agencies for the effective management of the medical resources during a Weapons of Mass Destruction event.

Over the past few years the Denver Field Office has developed an exceptional working relationship with the Colorado National Guard 8th Weapons of Mass Destruction Civil Support Team, which is stationed at the Buckley Air Force Base outside of Denver. This full time National Guard team is trained in field detection and assessment of chemical, biological and radiological materials, and has worked closely with FBI Denver's Hazardous Materials Response Team in the evaluation of numerous potential Weapons of Mass Destruction events. The 8th Weapons of Mass Destruction Civil Support Team is also working with the Denver Field Office to field advanced detection equipment in a pilot program which is unique within the FBI.

Working Groups

In December 1998, FBI Denver took a leadership role in the establishment of what is known as the Colorado Counter Terrorism Advisory Council. This group, which has met monthly since its inception, includes representatives from a variety of state and federal agencies who have primary responsibility for response to Weapons of Mass Destruction issues within the State of Colorado. In addition to the FBI its members include the Colorado Office of Emergency Management, the Colorado State Patrol, the Colorado Department of Public Health and Environment, the Colorado National Guard 8th Weapons of Mass Destruction Civil Support Team, the Environmental Protection Agency, the Federal Emergency Management Agency, the Public Health Service, the Department of Energy, and recently the Colorado Office of Preparedness, Security and Fire Safety. This group has established interagency notification and response protocols which have greatly enhanced our ability to efficiently share information and provide response resources. Based upon the success of the Colorado Counter Terrorism Advisory Council, a similar working group has been established in Utah and one is in the process of being established in Wyoming.

Operational Protocols

The Denver Field Office responded to its first anthrax threat letter in 1998. Numerous operational issues, including collection of hazardous evidence, field screening, and laboratory analysis, had not yet been fully developed. FBI Denver worked with its federal, state and local response partners to establish more efficient operational protocols, which were further refined by the continuing but sporadic threats and incidents that occurred from 1998 to September 2001. During that time, we enhanced our ability to evaluate the credibility of Weapons of Mass Destruction threats, established relationships with local, state and federal hazardous materials teams, and created agreements for laboratory testing of chemical, biological and radiological agents. FBI Denver also formalized our working relationships with the Colorado Department of Public Health and Environment and the Wyoming Department of Health, both designated as members of the Laboratory Response Network and our primary contacts for analysis of potential biological agents.

Since the deliberate distribution of anthrax in the United States Mail during September and October 2001, FBI Denver has responded to more than 800 telephonic requests for assistance concerning potential Weapons of Mass Destruction terrorism, primarily regarding anthrax. In addition, FBI Denver provided field responses to potential Weapons of Mass Destruction incidents on more than 100 occasions, and opened more than 30 criminal investigations. The ability to handle this many incidents was a direct result of the partnerships developed among the federal, state and local response community.

NATIONAL INFRASTRUCTURE PROTECTION CENTER (INFRAGARD AND KEY ASSET PROGRAMS)

In 1997, the President's Commission on Critical Infrastructure Protection published a report concluding that the interdependence of critical infrastructures and the reliance on new information technologies "have created a new dimension of vulnerability, which, when combined with an emerging constellation of threats, poses unprecedented national risk" and recommended the establishment of a new national structure to address the problem.

On February 26, 1998, The National Infrastructure Protection Center (NIPC) located at FBI headquarters in Washington, D.C. was established to address the cyber and physical threats to our Nation's critical infrastructures. The NIPC's mission is to detect, deter, warn of, respond to, and investigate malicious acts, both physical and cyber, that threaten or target the Nation's critical infrastructures. The NIPC is the national focal point for gathering information on threats to the infrastructure. These critical infrastructures are banking and finance, electric power, emergency services, government operations, transportation, and water supply systems. The NIPC is composed of representatives from the FBI; Departments of Defense, Energy and Commerce; the Intelligence Community; other federal agencies; state and local governments; and the private sector.

To carry out NIPC's mission the FBI developed the Key Asset Program (KAP) and the InfraGard Program (IP). The Denver Field Office of the FBI initiated the KAP in April 2000. The purpose of the program is to identify the key assets in the states of Colorado and Wyoming that are the critical infrastructures of this region and, if attacked, would result in a major loss or disruption to the critical services these infrastructures provide to the residents of Colorado and Wyoming or the Nation. The goal of the KAP is to enhance the cyber and physical security of these key assets by working with private industry and other federal, state and local governments.

The FBI also initiated the InfraGard Program. This program addresses the need for a private and public sector information sharing mechanism at both national and local levels. Individuals from private industry join local InfraGard chapters such as the one started by the Denver Field Office of the FBI in November 2000. Currently, the Denver Chapter has 75 members. By participating in local chapters members have access to an Alert Network to voluntarily report actual or attempted illegal intrusions, disruptions and vulnerabilities of information systems; can access a Secure InfraGard Web site with recent information about infrastructure protection; and can call the Help Desk at the NIPC to ask questions about the program. After the Terrorist Attacks on September 11, the FBI Denver InfraGard Chapter conducted or co-sponsored the following events to educate the community about critical infrastructure issues:

- Meeting concerning infrastructure matters with the United States Society of Dams
- Co-hosted the "Secure The Rockies" Conference with the University of Denver, National Law Enforcement and Corrections Technology Center. This was a 3 day conference attended by approximately 200 individuals from all sectors. Keynote speakers were Ronald L. Dick, Director, National Infrastructure Protection Center (NIPC), and Cyber Division Deputy Assistant Director James Farnan.
- The President's Critical Infrastructure Protection Board Town/Hall

meeting discussing the National Strategy to Secure Cyberspace. The keynote speaker was Richard Clarke, Special Advisor to the President for Cyberspace Security.

- Denver InfraGard Chapter quarterly meeting with the topic "The Economic Import of September 11 on Colorado" held at the University of Denver Law School. The keynote speaker was Sam Adomms, retired CEO of Frontier Airlines.

- Information Security Awareness Seminar for Banking and Finance Institutions

- FBI Denver/Colorado National Guard meeting with Key Assets to identify key assets and obtain input on assessing the vulnerabilities of sectors to cyber and physical attacks.

After the terrorist attack on September 11, the FBI consolidated its resources devoted to various programs related to infrastructure protection matters and computer crimes into a Cyber Crime Division at FBI headquarters with special agents assigned to cyber crime squads in field offices. The mission of the special agents assigned to these squads is to investigate violations of federal laws in which the internet, computer systems, or networks are the targets of terrorist organizations, foreign government sponsored intelligence operations or other criminal activity.

Denver's Cyber Crime Squad includes technically trained agents in the area of computer intrusion and a member of the FBI's Computer Analysis Response Team (CART). The CART member is responsible for computer forensic examinations. The CART representative also works jointly with the Colorado Regional Computer Forensic Laboratory (CRCFL) located in Douglas County. The CRCFL serves the law enforcement community by providing computer forensic examination services. By combining the CART and CRCFL resources the FBI is able to evaluate computer intrusions and other computer-related criminal activity in the State of Colorado. This squad also administer the Key Asset and InfraGard programs to better educate the local communities about threats to our critical infrastructures and how to assist the FBI in preventing future terrorist attacks.

CONCLUSION

Terrorism represents a continuing threat to the United States, and as the Lead Federal Agency for response to a Weapons of Mass Destruction threat or incident, the FBI must remain prepared to tackle this formidable challenge. In order to effectively and efficiently respond to the threat, the Denver Field Office of the FBI continues to enhance its Counterterrorism Program.

The Denver Field Office of the FBI retains its long-term commitment to working as a partner with state and local government in preparing to meet the challenge of a Weapons of Mass Destruction terrorist incident. Chairman Horn, this concludes my prepared remarks. I would like to express my appreciation to this subcommittee's examination of the issue of counterterrorism preparedness and I look forward to responding to any questions.

Mr. HORN. Well, I thank you, gentlemen, and thank you again for the good presentation. We now move to Peter Bakersky, Director of the Office of National Preparedness, Region VIII, Federal Emergency Management Agency, otherwise known as FEMA.

STATEMENT OF PETER BAKERSKY, EXECUTIVE OFFICER, NATIONAL PREPAREDNESS DIVISION, REGION VIII, FEDERAL EMERGENCY MANAGEMENT AGENCY

Mr. BAKERSKY. Good morning, Mr. Chairman. It's a pleasure for me to be here to discuss a pressing matter of how FEMA is assisting State and local government to prepare for a potential terrorist attack involving biological, chemical, or nuclear agents. FEMA is the Federal agency responsible for leading the Nation in preparing for, responding to, and recovering from disasters. Our success depends on our ability to organize and lead a community of local, State, and Federal agencies and volunteer organizations. The Federal Response Plan forms the heart of our management framework and lays out the process by which interagency groups work together to respond as a cohesive team to all types of disasters. In response to the terrorist events of 2001, the Federal Response Plan has proven to be an effective and efficient framework for managing all phases of disasters and emergencies. The plan is successful because it builds upon existing professional disciplines, expertise, delivery systems, and relationships among the participating agencies.

Much of our success in emergency management can be attributed to our historically strong working relationship with our State and local partners. Through our preparedness programs we provide the financial, technical, planning, training, and exercise support to give State, local, and tribal governments the capabilities they need to protect public health, safety, and property both before and after the disaster strikes.

In meeting the challenges ahead for State and local government, FEMA's Office of National Preparedness is becoming more robust. The mission of the Office of National Preparedness is to provide leadership in coordinating and facilitating all Federal efforts to assist State and local first responders, as well as emergency management organizations, with planning, training, equipment, and exercises.

FEMA has made the following changes to support this expanded mission. We have realigned preparedness activities from the Readiness, Recovery, and Response Directorate to the Office of National Preparedness. We have realigned all training activities into the U.S. Fire Administration to allow greater coordination between training for emergency managers and training for the first responders. We have moved the authority for credentialing, training, and deploying the urban search and rescue teams from the Readiness, Response, and Recovery Directorate to the U.S. Fire Administration.

We continue to work with all 55 States and Territories and federally recognized Indian tribes and Alaskan native villages to implement our current and other grant programs to assist State, tribal, and local governments to enhance their capabilities to respond to all types of hazards and emergencies, such as chemical incidents, incidents involving radiological substances, and national disasters.

We recognize that chemical, biological, and radiological scenarios will present unique challenges to the first responder community. Of these types of attacks, we are, in many ways, better prepared for a chemical attack because such an incident is comparable to a large-scale hazardous materials incident.

In such an event, the Environment Protection Agency and the Coast Guard are well connected to local hazardous materials responders, State and Federal agencies, and the chemical industry. There are systems and plans in place for response to hazardous materials, systems that are routinely used for both small and large-scale events. The EPA is also the primary agency for the hazardous materials function of the Federal Response Plan. We are confident that we would be able to engage the relevant players in a chemical attack based on the hazardous materials model.

Bioterrorism, however, presents the greater immediate concern. With a covert release of a biological agent, the first responders will be hospital staff, medical examiners, private physicians, or animal control workers instead of the traditional first responders, with whom we have a long-term relationships.

The Department of Health and Human Services leads the efforts of the health and medical community to plan and prepare for a national response to a public health emergency and is the critical link between the health and medical community and the larger Federal response. The Department of Health and Human Services is also our primary agency in the Federal Response Plan for health and medical services.

The Federal Radiological Emergency Response Plan has 17 Federal agency signatories, and the Nuclear Regulatory Commission is the lead Federal agency for coordinating the overall response, and FEMA is responsible for coordinating nonradiological support.

Tabletop exercises have been conducted in order to determine Federal agency resources for responding to a terrorist attack with a radiological component. In addition, nuclear or radiological threat posed by improvised nuclear devices and radiological dispersal devices is being evaluated, as well as the preparedness of member departments and agencies to deal with these threats.

It is FEMA's responsibility to ensure that the national emergency management system is adequate to respond to the consequences of catastrophic emergencies and disasters regardless of the cause. We rely on our partners at the State and local level. Without question, they need support to further strengthen their capabilities and their operating capacity.

FEMA must ensure that the national system has the tools to gather information, set priority, and deploy resources effectively. In recent years we have made tremendous strides in our efforts to increase cooperation between the various response communities and now we need to do more.

The creation of the Office of National Preparedness and our emphasis on training, planning, equipment, and exercises, will enable us to better focus our efforts and will help our Nation become better prepared for the future.

Thank you, Mr. Chairman, and I'll be available for any questions.

[The prepared statement of Mr. Bakersky follows:]

STATEMENT OF

PETER J. BAKERSKY

EXECUTIVE OFFICER

NATIONAL PREPAREDNESS DIVISION
REGION VIII

FEDERAL EMERGENCY MANAGEMENT AGENCY

BEFORE THE

SUBCOMMITTEE ON GOVERNMENT EFFICIENCY, FINANCIAL

MANAGEMENT AND INTERGOVERNMENTAL RELATIONS

COMMITTEE ON GOVERNMENT REFORM

U.S. HOUSE OF REPRESENTATIVES

AUGUST 23, 2002

EXHIBIT
7
HUNTER & GEIST

Introduction

Good morning, Mr. Chairman. I am Peter Bakersky, Executive Officer, National Preparedness Division, Region VIII of the Federal Emergency Management Agency (FEMA). It is a pleasure for me to be here today to discuss the pressing matter of how FEMA is assisting State and local governments to prepare for a potential terrorist attack involving biological, chemical or nuclear agents. I will describe how FEMA works with other agencies and our State and local partners, our programs related to terrorism, and new efforts to enhance preparedness and response. The President's proposal to create a Department of Homeland Security would strengthen these links and enhance our Nation's ability to respond to terrorism.

FEMA's Coordination Role

FEMA is the Federal Agency responsible for leading the nation in preparing for, responding to and recovering from disasters. Our success depends on our ability to organize and lead a community of local, State, and Federal agencies and volunteer organizations. We know whom to bring to the table when a disaster strikes in order to ensure the most effective management of the response. We provide management expertise and financial resources to help State and local governments when they are overwhelmed by disasters.

The Federal Response Plan (FRP) forms the heart of our management framework and lays out the process by which interagency groups work together to respond as a cohesive team to all types of disasters. This team is made up of 26 Federal departments and agencies, and the American Red Cross, and is organized into interagency functions based on the authorities and expertise of the members and the needs of our counterparts at the State and local level.

Since 1992, and again in response to the terrorist events of September 11, 2001, the FRP has proven to be an effective and efficient framework for managing all phases of disasters and emergencies. The FRP is successful because it builds upon existing professional disciplines, expertise, delivery systems, and relationships among the participating agencies. FEMA has strong ties to the emergency management and fire service communities and we routinely plan, train, exercise, and operate together to remain prepared to respond to all types of disasters. The National Strategy for Homeland Security proposes to build on the experience of the Federal Response Plan to develop one all-discipline, all-hazard plan to cover all events of National significance and clarify the roles and responsibilities of different levels of government.

State and Local Relationship

Much of our success in emergency management can be attributed to our historically strong working relationship with our State and local partners. Through our preparedness programs we provide the financial, technical, planning, training, and exercise support to give State, local and Tribal governments the capabilities they need to protect public health, safety and property both before and after disaster strikes. Our programs foster the partnerships that are so critical to creating a strong comprehensive national emergency preparedness system. Terrorism consequence management is just one component of our overall emergency management effort. For example, after September 11, Governor Ridge and Director Allbaugh agreed that there was a need to quickly assess State capabilities to effectively respond to acts of terrorism. FEMA assembled an interagency team with members from Department of Defense, Department of Education, Health and Human Services, Department of Justice and Environmental Protection Agency to visit the 50 States and territories to assess their readiness against 18 criteria and to identify priorities and shortfalls. We examined several categories such as critical infrastructure, personnel, plans, equipment and supplies communications and related capabilities. The results were provided in a classified report to Governor Ridge right before Thanksgiving.

Meeting The Challenge Ahead – Creating the Office of National Preparedness

On May 8, 2001, the President tasked the Director with creating the Office of National Preparedness within FEMA to "coordinate all Federal programs dealing with weapons of mass destruction consequence management within the Departments of Defense, Health and Human Services, Justice, and Energy, the Environmental Protection Agency, and other federal agencies." Additionally, the ONP was directed to "work closely with state and local governments to ensure their planning, training, and equipment needs are met."

The mission of the Office of National Preparedness (ONP) is to provide leadership in coordinating and facilitating all Federal efforts to assist State and local first responders (including fire, medical and law enforcement) and emergency management organizations with planning, training, equipment and exercises. By focusing on these specific areas, we can build and sustain our nation's capability to respond to any emergency or disaster, including a terrorist incident involving chemical, biological or nuclear weapons of mass destruction and other natural or manmade hazards.

FEMA has made the following changes to support this expanded mission to support the Office of Homeland Security:

- Realigned preparedness activities from the Readiness, Response and Recovery Directorate to ONP;

- Realigned all training activities into the U.S. Fire Administration to allow greater coordination between training for emergency managers and training for firefighters;

- Moved the authority for credentialing, training and deploying Urban Search and Rescue teams from the Readiness, Response and Recovery Directorate to the U.S. Fire Administration.

ONP Organization

The ONP is organized in FEMA Headquarters under a Director (reporting directly to the FEMA Director) and supported by a Management Services Unit and four Divisions to carry out key its functions to coordinate and implement Federal programs and activities aimed at building and sustaining the national preparedness capability. The divisions and their functional responsibilities include the following:

- **Administration Division** – Provide financial and support services, and management of the grant assistance activities for local and State capability building efforts.
- **Program Coordination Division** – Ensure development of a coordinated national capability involving Federal, State, and local governments, to include citizen participation, in the overall efforts to effectively deal with the consequences of terrorist acts and other incidents within the United States.
- **Technological Services Division** – Improve the capabilities of communities to manage technological hazard emergencies- whether accidental or intentional-and leverage this capability to enhance the capability for dealing with terrorist attacks.
- **Assessment and Exercise** – Provide guidance, exercise, and assess and evaluate progress in meeting National goals for development of a domestic consequence management capability.

We continue to work with all 55 states and territories and Federally recognized Indian Tribes and Alaskan Native Villages to implement our current and other grant programs to assist State, Tribal and local government to enhance their capabilities to respond to all types of hazards and emergencies such as chemical incidents, incidents involving radiological substances, and natural disasters.

The Approach to Biological and Chemical Terrorism

We recognize that biological and chemical scenarios would present unique challenges to the first responder community. Of these two types of attacks, we are, in many ways, better prepared for a chemical attack because such an incident is comparable to a large-scale hazardous materials incident.

In such an event, EPA and the Coast Guard are well connected to local hazardous materials responders, State and Federal agencies, and the chemical industry. There are systems and plans in place for response to hazardous materials, systems that are routinely used for both small and large-scale events. EPA is also the primary agency for the Hazardous Materials function of the Federal Response Plan. We are confident that we would be able to engage the relevant players in a chemical attack based on the hazardous materials model.

Bio-terrorism, however, presents the greater immediate concern. With a covert release of a biological agent, the 'first responders' will be hospital staff, medical examiners, private physicians, or animal control workers, instead of the traditional first responders such as police, fire, and emergency medical services, with whom we have a long-term relationship. Across the Government, we are working to enhance our ability to detect biological attacks, better link the public health and emergency response communities, and train and equip traditional first responders to respond to bioterrorism. The President's proposal to create a Department of Homeland Security would strengthen the linkages, detailed below, that are critical to our capacity to respond to bioterrorism.

In particular, FEMA has worked with DHHS for several years on the Metropolitan Medical Response Systems (MMRS), which brings together various local medical response elements that have effectively planned, trained and prepared to respond to treat victims of mass casualty events, including chemical, radiological and biological terrorism. Under the program, participating cities plan for the equipment, supplies, training, and transportation requirements for emergencies including possible terrorist attacks.

In exercise and planning scenarios, the worst-case scenarios begin with an undetected event and play out as widespread epidemics, rapidly escalating into a national emergency. Response would likely begin in the public health and medical community, with initial requests for Federal assistance probably coming through health and medical channels to the Centers for Disease Control and Prevention (CDC).

DHHS leads the efforts of the health and medical community to plan and prepare for a national response to a public health emergency and is the critical link between the health and medical community and the larger Federal response. FEMA works closely with the Public Health Service of DHHS as the primary agency for the Health and Medical Services function of the Federal Response Plan. We rely on the Public Health Service to bring the right experts to the table when the Federal Response Plan community meets to discuss biological scenarios. We work closely with the experts in DHHS and other health and medical agencies, to learn about the threats, how they spread, and the resources and techniques that will be needed to control them.

By the same token, the medical experts work with us to learn about the Federal Response Plan and how we can use it to work through the management issues, such as resource deployment and public information strategies. Alone, the Federal Response Plan is not an adequate solution for the challenge of planning and preparing for a deadly epidemic or act of bioterrorism. It is equally true that, alone, the health and medical community cannot manage an emergency with biological causes. We must work together.

In recent years, Federal, state and local governments and agencies have made progress in bringing the communities closer together. Exercise Top Officials (TOPOFF) 2000 conducted in May 2000 involved two concurrent terrorism scenarios in two metropolitan areas, a chemical attack on the East Coast followed by a biological attack in the Midwest. This was a successful and useful exercise and we continue to work to implement the lessons learned.

In January 2001, the FBI and FEMA jointly published the U.S. Government Interagency Domestic Terrorism Concept of Operation Plan (CONPLAN) with DHHS, EPA, and the Departments of Defense and Energy, and these agencies have pledged to continue the planning process to develop specific procedures for different scenarios, including bioterrorism. The Federal Response Plan and the CONPLAN provide the framework for managing the response to an act of bioterrorism, but we need to continue to practice our response to events of this kind.

The Approach to Nuclear Terrorism

There are 63 commercial nuclear power plant sites in the United States, located in 33 States. These states and their local governments have radiological emergency response plans for the 10 miles surrounding the plants and 36 states have plans for the 50 miles radius surrounding the plants.

The Federal response to a nuclear power plant incident is documented in the Federal Radiological Emergency Response Plan (FRERP), which has 17 Federal agency signatories. The Nuclear Regulatory Commission (NRC) is the lead Federal agency for coordinating the overall response and FEMA is responsible for coordinating non-radiological support.

The FEMA Radiological Emergency Preparedness (REP) Program also routinely tests and evaluates the individual site plans, the 10-mile plans for the 63 sites are tested at biennial exercises (approximately 32 exercises per year) and the 50-mile plans for the 36 States are exercised once every six years (approximately six exercises per year).

The events of September 11 have now horrifically demonstrated that these plans needed to be expanded further. When September 11 showed us how a commercial jetliner can be used as a weapon of mass destruction, the NRC and FEMA began to work jointly on the preparation of protocols and procedures for dealing with the consequences of a similar attack on a nuclear power plant – a scenario previously not addressed. While some amendments to the emergency response plans may result from this review, it is important to note that the current plans are a valid approach to any nuclear power plant incident, regardless of the cause: terrorism, human error, technological failure, or a natural hazard.

The Federal Radiological Preparedness Coordinating Committee (FRPCC) has also conducted tabletop exercises of the FRERP in order to determine Federal agency resources for responding to a terrorist attack, or multiple attacks, with a radiological component. In addition, the FRPCC is evaluating the nuclear/radiological threat posed by Improvised Nuclear Devices and Radiological Dispersal Devices and the preparedness of FRPCC member departments and agencies to deal with these threats.

In addition, the Federal Response Subcommittee of the FRPCC has developed information on radiological terrorist devices--such as radiological dispersion devices, improvised nuclear devices, and radiological exposure devices--for the use of the Federal Bureau of Investigation as background and public information.

At FEMA, the creation of the Office of National Preparedness and our emphasis on training, planning, equipment, and exercises will enable us to better focus our efforts and will help our nation be better prepared for the future. The President's proposal to create the Department of Homeland Security will integrate these capabilities into a broader whole that will help our Nation respond to the terrorist threat.

Thank you, Mr. Chairman. I would be happy to answer any questions you have.

Mr. HORN. Thank you. And now we have Dr. Miller, the State epidemiologist for bioterrorism, Colorado Department of Public Health and Environment.

STATEMENT OF DR. LISA A. MILLER, STATE EPIDEMIOLOGIST FOR BIOTERRORISM, COLORADO DEPARTMENT OF PUBLIC HEALTH AND ENVIRONMENT

Dr. MILLER. Thank you, and good morning, Mr. Chairman. The previous speaker, I think, gave me a great lead-in to talk about the ways that the Federal Government, and specifically the U.S. Department of Health and Human Services, is aiding both State and local public health and hospitals to become better prepared to deal with bioterrorism.

The Colorado Department of Public Health and Environment is the recipient of two Federal grants recently to help us become better prepared. The first grant I'm going to just touch on briefly is Colorado's hospital preparedness grant. And these grant funds are provided through the Department of Health and Human Services via the Health Resources and Services Administration, and they're intended to upgrade the preparedness of hospitals. This grant was developed by the department, but there is oversight from a committee, and this committee has a broad range of representation. And I think that's a really important point to make.

This is obviously a very complicated area, and I think you've gotten the picture just from the few speakers before me that we have a lot of coordination to do. So to help coordinate this grant, we have an advisory committee from Veterans Affairs, Indian health service, from community health agencies, and from private providers, to really give us that additional perspective. The Hospital Preparedness Advisory Committee is led by Mr. Larry Wall, who is going to give you comments later. So I'm not going to go into any more detail about that grant, and I'll let Mr. Wall take that one further.

The second grant I want to mention briefly is Colorado's public health bioterrorism preparedness grant. And this grant is intended to build public health infrastructure, both at the State public health level and at the local agency level. This grant is actually about 3 years old. We started receiving Federal funds in public health for bioterrorism preparedness in 1999, but recently the Federal Government has increased those funds dramatically, as I'm sure you're aware. We went from receiving about $1 million a year to, this year, $14.6 million.

And I want to point out a couple things about this grant. First of all, although the grant is called a bioterrorism grant, it really is intended to help improve the infrastructure of public health so we can respond not only in the, hopefully, unlikely event of a major bioterrorism event, but we can respond to everyday emergencies which we have in public health all the time. We have an example right now going on. We have a new disease in our country, West Nile virus. So this is intended really to help us respond to both that, "everyday emergency" in public health and the other bigger issue of a bioterrorism event.

Again, in this grant, we both wrote the grant and have oversight in this grant by a committee. And we wrote the grant with the input of both a broad range of State health department individuals and local health department individuals. And then we have an advisory committee, and that advisory committee is actually made up of several individuals in this room. Ms. Mencer sits on the advisory committee. Major General Whitney, Mr. Wall, Mr. Greer and Agent Airy have been involved in the work of the committee. So, again, we are trying to really get some coordination between our different agencies and make sure that the one knows what the other is doing and has input into those activities.

The grant, as you see in my notes, is divided into six focus areas. These focus areas are funded separately. They deal with specific readiness preparations, that is, the writing of the plans, the exercising of those plans, and training. They also deal with surveillance and epidemiology, which is the ability to detect and respond to an event quickly. So if there is a bioterrorism event, we will know about it as soon as possible, and we can control it quickly and prevent spread.

There's also a laboratory part of this grant that will go specifically to help us upgrade our laboratory capacity so we can better identify, more quickly identify, agents of bioterrorism at the State level. We also want to upgrade local laboratory capabilities.

There's also a section that deals with information technology, obviously an important issue that runs through every part of terrorism prevention and response. And we are dealing not only with actually connecting people and making sure they have good communication technology, but protecting that information and making sure that our information is secure.

We found out with the anthrax event, I think, that risk communication or the ability of public officials and public health officials to communicate information quickly and accurately, is very important and can really mean the difference in an event between quick control and not-so-quick control of an event. We need to make sure that we have better training in risk communication in our public health and public official levels so that we can provide information in the best way possible to the public.

The last area of focus in our grant is training and education. Obviously we have a huge number of not only public health people but also hospital staff to whom this is a relatively new issue, and we have a lot of training to do about the issue, about how we respond to this issue, importantly, how we work with all our different partners here, which public health staff are not as accustomed to working with as it should be, probably. We are getting better at working with our partners and knowing who those partners are, but that's still an area of need to do some training and education in.

My final point that I would make with these funds is that they are intended to increase our infrastructure in public health. And I think in order to do that, a 1-year grant is not going to work. We need to hire personnel and have training that is very long-term. And to do that well, I think these funds, obviously, need to be longer than 1 or 2 years to actually improve our infrastructure in a long-term way. So I thank you for your time and attention, and I'd be happy to take questions.

[The prepared statement of Dr. Miller follows:]

Dr. Miller

8/21/2002

Summary

Public Health Preparedness and Response for Bioterrorism Colorado Department of Public Health and Environment

Federal Funding from the
Centers for Disease Control and Prevention and Health Resources and Services Administration
U.S. Department of Health and Human Services

Colorado's Grant Applications for Public Health Bioterrorism Preparedness and Hospital Preparedness were prepared in response to President George W. Bush's initiative and federal legislation that provided funds to foster bioterrorism preparedness among state and local health departments and hospitals. In a January 31, 2002 letter, Secretary Tommy G. Thompson directed each state governor to review and approve the state's applications, which was done prior to the submission of the applications in April, 2002. The content of each application is discussed in further detail below.

Colorado's Hospital Preparedness Grant

The Hospital Preparedness grant was developed by Colorado Department of Public Health and Environment with oversight by a Hospital Preparedness Advisory Committee. The Committee consists of representatives from the fields of emergency medical services, emergency management, rural health, veterans affairs, Indian health, public health, providers, the Metropolitan Medical Response System, law enforcement, and is chaired by the President of the Colorado Health and Hospital Association.

These grant funds are provided through the Department of Health and Human Services, Health Resources and Services Administration (HRSA) to upgrade the preparedness of Colorado's hospitals and collaborating entities (pre-hospital/emergency medical systems and community health clinics) for bioterrorism response.

Hospital Grant Award

The State was awarded $1.9 million, which is being delivered in two phases: The Phase One funds ($383,267) were awarded in March, 2002 to develop the infrastructure of the program, including the Hospital Preparedness Advisory Committee and a hospital needs assessment survey. The grant supports 1.75 staff positions (1.0 staff position for a program director, 0.25 staff for a medical director, and 0.50 staff for an administrative assistant); 54% of the Phase 1 funds will be combined with the Phase 2 funds and distributed directly to hospitals and other entities. The Phase two funding ($1.5 million) is expected to arrive in late September. More than 80% of these funds will be distributed directly to the hospitals and other entities, based on the identified needs from the Phase One needs assessment. The Hospital Preparedness Advisory Committee will prioritize the

EXHIBIT
8
HUNTER & GEIST

identified needs and items will be purchased in bulk, where possible, then distributed to hospitals as appropriate. These funds cover the time period up to March 31, 2003.

Hospital Grant Criteria

The grant criteria require: 1) identification of the needs of hospitals and other entities for bioterrorism readiness; 2) development of regional plans to accommodate a minimum of 500 victims, including identification of pharmaceutical needs, identification of specialty care patient needs (pregnant women, children, immuno-compromised persons, and elderly persons); and 3) provision of training. To maximize funding, this program is working in collaboration with the Public Health Bioterrorism Preparedness grant staff and with other agencies throughout the State that are developing bioterrorism response plans with other federal funding sources.

Colorado's Public Health Bioterrorism Preparedness Grant

Colorado's plan was developed by a Planning Group comprised of 10 staff from the Colorado Department of Public Health and Environment and 11 representatives from local health departments and local public health nursing groups, with input and oversight from a multi-disciplinary Bioterrorism Advisory Committee. The Committee was formed by supplementing the Governors' Expert Emergency Epidemic Response Committee with additional representatives from the areas of mental health, occupational health, emergency medical services, rural health, police and fire department workers, and community health centers.

The federal grant funds available for public health preparedness total $14.6 million for the time period June 2002 through August 2003. These funds are apportioned by formula to six focus areas, summarized below. Each focus area also contains one or more "Critical Benchmarks" which are activities that the Department was required to complete at the time of the grant submission (April 15, 2002). Among these benchmarks was the appointment of Dr. Ned Calonge to serve as the Executive Director of the State Bioterrorism Preparedness and Response Program. The federal public health bioterrorism grants were awarded to each state and several cities in the nation using a base plus per capita formula. The grants are intended to build public health infrastructure so that the public health system will be better able to respond to bioterrorism and other public health threats and emergencies.

Of the funds allotted, a total of $2.3 million has recently been made available to county and regional public health agencies in the state to provide immediate funds for addressing issues of training and preparedness planning and response.

8/21/2002

Focus Area A: Preparedness Planning and Readiness Assessment
$3.3 million

Grant criteria involve: 1) assessing current capacity and developing, enhancing and maintaining public health leadership, expertise and workforce necessary to address bioterrorism, other outbreaks of disease, and other public health emergencies; 2) developing and testing coordinated preparedness and response plans, and 3) developing and testing a plan for deployment of the National Pharmaceutical Stockpile.

Colorado funds will be used to fund 11 new staff in local health departments and related regions for planning and exercises, coordinated with five staff (2.5 new, and including a currently funded position in the Colorado Office of Preparedness, Security and Fire Safety) at the state level. Additional funds will support preparedness assessment, planning and exercise activities, and equipment purchases such as personal protective equipment necessary for response to a bioterrorism event. Statewide and regional multidisciplinary conferences are planned (the first is occurring October 7 and 8, 2002) to provide training to a broad constituency. Approximately $1.7 million is provided to local public health agencies.

Focus Area B: Surveillance and Epidemiology Capacity
$3.0 million

Grant criteria involve developing, enhancing, supporting and testing the epidemiologic infrastructure to 1) rapidly detect a terrorist event, 2) rapidly and effectively investigate, respond to and control the consequences of a terrorist event, and 3) respond to other naturally occurring disease outbreaks and public health threats (to provide evidence of preparedness). Additional criteria involve strengthening working links between responding areas (public health, health care and law enforcement).

Colorado funds will be used to fund 13 new epidemiologists at the local level for outbreak and event detection, investigation, response and consequence management, coordinated with four staff (two new) at the state level. Additional funds will support improvements to existing reporting systems, development of new surveillance systems, and support other activities to meet grant criteria. Approximately $1.5 million is provided to local public health agencies.

8/21/2002

Focus Area C: Laboratory Capacity—Biologic Agents
$1.9

Grant criteria involve 1) developing a state-wide laboratory services program to support response to bioterrorism, other infectious disease outbreaks, and other public health threats and emergencies and 2) ensure adequate laboratory infrastructure to identify potential biologic agents from a bioterrorism event.

Colorado funds will be used to improve communication between public health and other types of laboratories, improve the level of six local public health department labs to detect biological bioterrorism agents, and to expand the capabilities of the state laboratory. Laboratory security features will also be improved. Seven positions are funded in this area, including 3.5 new staff to support coordination, training and expanded capabilities. Approximately $500,000 is provided to laboratories at the local level.

Focus Area E: Health Alert Network/Communications and Information Technology
$1.7 million

Grant criteria involve 1) insuring connectivity among event participants with high-speed Internet access, 2) developing a method of emergency communication for event participants, 3) protecting data and information systems, and 4) securing electronic exchange of public health information.

Colorado funds will be used to provide high-speed Internet access to all local public health venues, to provide Health Alert Network access to the same venues, and to evaluate, purchase and implement a communication system for use in the event of a bioterrorism attack. Funds will also be used to assure the security of our communication system. Three new state staff will join 0.5 existing staff to meet these goals. Approximately $900,000 is provided to local health agencies to support connectivity and communication needs.

8/21/2002

Focus Area F: Risk Communication and Health Information Dissemination (Public Information and Communication)
$0.8 million
Grant criterion is to develop the resource to provide adequate, appropriate and timely health risk information to the public in response to a terrorist event or other public health emergency.

Colorado funds will be used to train public health professionals in the area of risk communication, to determine the appropriate methods of communication to reach specific populations in Colorado, to develop statewide and local risk communication plans, and to implement those plans. This will involve one new staff at the state level to coordinate this project. Approximately $430,000 is provided to local health agencies in this area.

Focus Area G: Education and Training
$1.5 million

Criteria involve 1) ensuring the delivery of education and training to appropriate participants in a bioterrorism event, 2) ensure that such participants in the community are identified and could be mobilized in response to and event, and 3) evaluate the effectiveness of training through drills, simulations and actual event responses.

Colorado funds will be used to support a system of organized training opportunities in person and through distance learning methods. Curricula will be taken from national sources when possible and developed where needed. Funds will support implementation of a system to assess and track ongoing training needs statewide. The grant will fund four local health department/regional staff to support and coordinate local training needs and activities, coordinated with three new staff training coordinators at the state. Approximately $330,000 is provided to local agencies in this focus area.

Mr. HORN. Well, you've done a good job because we hadn't had a chance to get into these grants since they have been trickling out over the last few months, and that's very helpful.

So now let's move to the question and answer. What I'm going to do is have each of us, my colleague and myself, each have 10 minutes for questions and answers. And if there's still questions and answers to be had, we will do it again. So we'll start with my colleague here. You have 10 minutes.

Mr. TANCREDO. Thank you, Mr. Chairman. You know, I've got a couple of things that have come to mind as I listened to the testimony. And I, again, thank you all very much for a clear, concise presentation.

Major General Whitney, in terms of the new role, the added role, perhaps I should say it that way, for the Guard and specifically in the areas that you mentioned in terms of airport security and that sort of thing, how has that affected, if at all, the morale of the people involved in terms of a change in what they see, I suppose, as their primary task or role? I mean, you know, other than—I mean, it seems to me that up to this point in time they saw themselves in a much more active role in case of an emergency, getting in and helping people, rescuing people, doing all the things that the Guard has been so good at doing, rather than standing there, you know, for hours and hours and hours and hours looking at people walking past them. And I just wondered how, if anything, that has affected the actual morale. Do you have any indicator of that?

General WHITNEY. Sir, I can answer that fairly specifically as well as generally. No. 1, a general answer to the question is that, of course, we've been providing homeland security since 1636 throughout the Nation, so we are not—this is not a mission that is totally new to us. So we have done these types of things more than just the national defense mission where you have a soldier who is well versed in operating field artillery is all at once asked to come in and provide airport security. A trained military force is a very capable force in many different missions. Of course they're trained in combat arms, and so therefore airport security missions in support of law enforcement organizations would be something that would fit within their skill set.

The specific answer to your question, though, how has it affected the morale of the soldiers who are involved in that mission is, it hasn't affected it adversely whatsoever. Actually, it's been a very good thing for most of our soldiers with very few exceptions. All of our soldiers who performed that mission did so with a very, very positive attitude about representing the U.S. military, specifically the Colorado National Guard and the U.S. Army, in a very visible manner in that airport security mission. So they're very proud to do that, very proud to stand in their uniforms providing that added sense of security that would come as a result of their presence.

Mr. TANCREDO. That's interesting to know. Just as I—Every single week as Mr. Horn, I'm sure, and I are required to do for the job and fly in and out of Denver twice a week and passing them each time, I always just thought to myself, "I wonder if they just get so damn bored that they can't stand this anymore." And "How do you stay alert?" And "How does it"—that's a challenge, I think, and God bless the Guard for the good job they do. But it just always made me wonder about that.

Ms. Mencer, the focus of most of the discussion here today was naturally on the reactive capacity of the State in case of an emergency. But you mentioned something that piqued my curiosity when you started talking about what I would have categorized, I suppose, as a preventive activity or responsibility that you may find in your own—because it's the office of preparedness, you're preparing for something as opposed to trying to prevent it from happening.

So when you talk about the information sharing and that sort of thing, I just wonder to what extent do you actually see your role, the role of your agency, in this other capacity of preventive, and how specifically does that play out, if at all? What do you do in that regard?

Ms. MENCER. Well, certainly at the State level we are not involved in investigations concerning terrorism. That would be the responsibility and purview of the FBI. But I think what we've learned as a Nation is that we need to have a better mechanism for collecting intelligence and to share it, and we are hoping that at the State level, by working with the local chiefs and sheriffs, we can have a mechanism for doing that, for going out and looking for things that they would have an interest in, and then being able to disseminate that information to the right people.

We'd like to go beyond that, though, rather than just law enforcement, because I think, as Dr. Miller said, we have a lot of health workers out there that see things every day. We have a lot of first responders that arrive on the scene first, and they need to have some intelligence as well. So we are trying to develop a system where we can not only disseminate bulletins that come out from the FBI at a law enforcement level, but the Office of Homeland Security is actually looking at trying to have a tear sheet so that information can be scrubbed and disseminated to first responders as well, to the health care workers, to the first emergency responders who come on the scene, to the fire chiefs, so that they have a way of knowing what's going on too. So if there is an explosion, they might know that, gee, we've had pipe bombs discovered in other parts of the State. Maybe it's a criminal act rather than just a gas explosion.

So those are the kinds of things we are looking at trying to facilitate. It's something that we haven't done well before, we haven't done at all in some cases, particularly with first responders. So it's a new way of looking at intelligence, and we are hoping to play a part in that to assist the investigators in their role with looking at what do we have in this Nation.

Mr. TANCREDO. Well, that's still sort of the reactive stage which you've just described, how do we react to the event, how quickly can we get the information to the people that will be there first to know what they're dealing with.

But you suggest that this information sharing task that you've been charged with is the primary role of the preventive side of your activity. And so I guess I'm asking you to be even more specific, if you could, and exactly, No. 1, is that happening at all today? You mentioned, I think, that you are looking to ways in which that can occur. Is it happening now? Are you facilitating that and to what extent do you think that we have increased that degree of information sharing as a result of the agency's existence, your agency's existence? How much more do you look forward to doing that in that regard?

Ms. MENCER. Indeed we are doing it today, and we started shortly after the office was formed. Right now we are cobbling together several different communications systems to disseminate information. So we now use our CCIC, our criminal system that we have in place, to disseminate bulletins and law enforcement sensitive information. We use the Colorado Law Enforcement Information Network, or CLEIN, to disseminate information as well. We use RISSNET, which is an Internet-based system to disseminate information. We also use an e-mail system that's developed by my Director of Fire Safety that disseminates information that isn't law enforcement sensitive to fire chiefs and emergency first responders. So we indeed send this out, and we haven't done that before. So that has been something new. We've received positive comments from local law enforcement agencies and from first responders, that for the first time they're actually getting some kind of intelligence information.

So that's been working well, but we need a better system instead of trying to use all these different systems, and we are working with the RISSNET people in a system that the Colorado State Patrol already has in their possession, Dialogic, which is a communicator system. What we found with chiefs and sheriffs is if their information comes in via the Internet or via their teletype machine or whatever system they have, they don't know that it's there. There's no mechanism with their own business practices for the clerical person when he receives it to say, "Wow, this is something significant; the chief needs to see it." So a lot of times it sits there. So the Dialogic system, if you plug your numbers in, your fax number, your pager, your cell phone, your home number, it automatically begins to call people and says, "You need to check your fax machine, your Internet message," whatever. And so then they know. And it keeps calling them until they respond. So it's a very annoying system, but it works.

And so we are looking at using the money from the U.S. attorney's office, which they have been provided, and we have requested some funding to hire a person that will do nothing but operate this Dialogic system so that we do have a communication effort where we can alert people that there is a message coming in and they need to respond to it.

Mr. TANCREDO. And from what agencies are you receiving most of the information that you are presently then disseminating?

Ms. MENCER. We get information from the FBI on their intelligence bulletins that they disseminate weekly. We also get the same information——

Mr. TANCREDO. Excuse me. Now, those bulletins would not have otherwise—let's assume that your agency didn't exist for a moment—they would not have otherwise been disseminated?

Ms. MENCER. They do. They are disseminated on our CCIC system, which I think would have occurred anyway. Also the NLET system, which the FBI uses, and we use that as well.

Mr. TANCREDO. So that now that is not happening that way; you sort of have taken over that?

Ms. MENCER. Well, no, they—Do you disseminate on NCIC? I don't know if you do or not. I think it's CBI now.

Mr. CARBALLIDO. Well, we do from headquarters, initially, and then there's further dissemination from CCI, and sometimes we also disseminate from FBI Denver BOLOs, be on the lookout for this, etc., more specific operational tactical information, and that goes directly to the local law enforcement agencies.

Mr. TANCREDO. What I'm trying to figure out here is exactly what the role is in term of this information dissemination for your agency. And is it truly a coordinating agency or are you just another part of the dissemination picture? I'm not clear on that exactly.

Ms. MENCER. I think it's both. I think what we've done is create this whole other layer for first responders and for the fire chiefs that didn't exist before. And they had no information flow to them at all. So, you know, that's something that we've created in the new office and we are trying to enhance. And if the Office of Homeland Security proceeds with their goal to have that tear sheet, it will be much facilitated because they'll provide scrubbed information, if you will, on the bottom part that we can disseminate to first responders and then the law enforcement sensitive stuff at the top. So we are hoping that will make our job a lot easier as well.

Mr. TANCREDO. So, so far, it's mostly sort of—right now would you consider it to be a top-down information sharing process? You are not getting information, let's say, from sheriffs' departments that you then—coming up to you that you—I guess it's not—I don't know whether it's correct to say coming up to or down from, but are you getting it and are you sharing the information that they provide to you?

Ms. MENCER. That's what we need to work on specifically, and that's what the seven districts will do, and the CBI component of the those seven districts will work with the chiefs and the sheriffs in collecting information that then we can disseminate statewide. For instance, if District 1 says, you know, we've received some information that's some driver's licenses have been stolen, and then we'll be able to send that out to the seven districts, and we'll coordinate the intelligence, bring it to the FBI's attention so that they can see is this something we need to be concerned about or not.

So as we work on this process of getting these districts organized and reaching out to the chiefs and the sheriffs, we'll have a method of not only disseminating it from the top down, but then disseminating it from the bottom up, which we need to do and which we haven't done very well as a Nation, I don't think.

Mr. TANCREDO. Well, I appreciate the chairman's indulgence here on this because I'm just trying to get a good, clear picture of exactly what it is that you are charged with doing and how effectively it has, you know, begun to operate so far and where you think you want to go with it.

It is confusing because there is this, you know, sort of a generic application of the word "dissemination," and we use it a lot, and we talk about the need for information sharing, and are never sure if just the creation of an agency that is supposed to help in that purpose can actually be that single source or it just becomes another one of the things that people sort of look forward to hearing from periodically, people sort of down the line from it. And it's a challenge, I think, enormous challenge, of course, to figure out exactly what role you're going to play that isn't being played by some other—by the FBI and other Federal agencies.

But let me take it one step further then. To what extent do you communicate or have communications with other Federal agencies that have responsibilities for internal security, specifically, let's say, the INS, Customs, even more specifically, Border Patrol; perhaps that wouldn't be the one because, although they do have, of course, internal security apparatus, to a couple of agents, anyway, do you have any involvement with them whatsoever?

Ms. MENCER. Well, we are a part of the CTAC, which was mentioned earlier, the counterterrorism committee that meets, the State patrol, so we do have participation with other Federal agencies on that. I'm in regular communication with ASAC Carballido, so I speak on a frequent basis with the FBI because I have a relationship with them, obviously, which I think is a good thing at this point in our Nation's history, particularly in this State that I do have that relationship with the Bureau, and I appreciate that.

And we, as a State, I don't see our role as coordinating with Federal agencies as much as I do with trying to coordinate the local folks out there. And I think because we in law enforcement are particularly territorial with the information we collect and receive, that we do need to begin to share more with sheriffs, sharing with the chiefs, and chiefs sharing with the sheriffs, and that kind of thing, and I see the State's role more as that coordination piece. There is an inherent distrust, I think, in law enforcement between local police agencies and the Federal agencies, and I think that is historic and has been going on for a long time. I don't think we are about to change that quickly. But I don't see the Federal Government being able to come in and do that. I think at a State level it's much easier for us to say we all need to work together in this, and then help with that intelligence piece to provide information to the FBI. But, as Mr. Carballido and I have discussed, it needs to go both ways, and the FBI needs to begin to bring things down as well, and hopefully we'll act as a conduit for that too.

Mr. TANCREDO. Maybe the creation of the Homeland Security will help—hopefully that will help in this rather confusing and sometimes convoluted process.

And the last question I have is for Dr. Miller. There is an agency I visited sometime ago, and I apologize because it slips my mind, but it is located at Buckley. It's unique. Maybe I shouldn't say unique, but there may not be more than one or two others around the Nation. It has the ability to respond immediately to an event, chemical, biological type of event, and tell the State exactly what it is with which we are dealing with that particular agent. And I'm just wondering if—I didn't hear anything about it, I guess, or any coordination of activity with that. It seems like it was an enormously—When I visited, I thought to myself, "God, what a great asset to have here in Colorado," because I think it's——

Dr. MILLER. Are you talking about the civil support team?

Mr. TANCREDO. That's it.

Dr. MILLER. That's General Whitney's. And, actually, we work very closely with them. If they're evaluating, say, an anthrax threat, we are often talking to them. And our lab serves as a confirmatory lab. If their testing showed that they thought they had some agent, our lab would test it and confirm it. So we are very familiar with them.

Mr. TANCREDO. General, maybe just to share with the committee what the responsibility—because I thought that was just so great when I was there, I thought, you know, this is in Colorado only or very few States, anyway.

General WHITNEY. Well, no, sir; actually there are 27 teams right now. We were the very first team organized, very first team certified. We've got the best team in the Nation here in Colorado. I'm not biased as all in that.

Mr. TANCREDO. Tell us exactly what its responsibility is.

General WHITNEY. It's a team—The designation is Weapons of Mass Destruction Civil Support Team, and we are team No. 8, is our designation, because of FEMA Region VIII as to what we were originally assigned to.

Specifically, they are designed to take this 22-person team to a site that has been designated as a possible nuclear, biological, chemical, radioactive, or even a high explosive site, to determine if there is any type of agent there that requires special protective measures as well as to mitigate whatever it is that they find. They have tremendous reach-back capability with a communications suite, a truck that has every kind of communications gear you can think of. They have a tremendous capability to analyze on the scene with a mobile analytical laboratory, which is probably what you saw when you went out to Buckley.

It's a one-of-a-kind-in-the-nation capability for this team, obviously, for the 27 teams that have this. So it's a tremendous asset for any community to have in place already, but it's also an asset that can be deployed anywhere in the Nation, whenever the need arises.

Mr. TANCREDO. Again, I really appreciate the chairman's indulgence here.

Mr. HORN. No, I think when you've got a good topic, keep going.

Mr. TANCREDO. Well, that's it. I guess I'd say that the one thing I remember about that tour is that there was a hope on their part that there would be a lot more knowledge, general knowledge, about their existence and about their ability to actually coordinate than evidently was the case then. They felt that they were somewhat unused, that there was—you know, here we are, this great thing, and how many people even know, how many even local agencies know that we are there to respond if they have something like that. So that's why I kind of wanted to bring it out here, and, hopefully, I don't know, just get people to be more aware of its existence which, again, seemed like a great asset.

General WHITNEY. Sir, I'm not sure how long ago you visited the organization out at Buckley.

Mr. TANCREDO. It could have been a week ago or 2 years ago, in my mind, I don't know; they all kind of fall together.

General WHITNEY. Within the last 8 months or so, they have been very active and responded to numerous calls in support of the FBI as well as local agencies. So I think we have done, I think, a much better job in advertising their capabilities to all those responders that may require their capability.

Mr. TANCREDO. Thank you, General. Thank you, Mr. Chairman.

Mr. HORN. Let me followup in another way involving Ms. Miller and Ms. Mencer and also the general, and that is, what do we have in hospital capacity should there be some attack of biology, chemistry, nuclear, whatever? And I just wonder to what degree and I guess we will address that in the other panel and get a lot of knowledge there. But what's your figure on that?

Dr. MILLER. Well, actually, surge capacity in our hospitals is a huge issue and there are few empty beds right now, so this is a major problem in the case of a bioterrorism event or any other event that requires a lot of hospital beds. And one of the major objectives of this hospital grant is to create regional hospitals that will serve 500 patients, which is far beyond what we could do today. So we need to be creative and think of ways and places where you could actually take care of 500 patients in a region if you had to.

And people are also looking at ways to increase bed capacity in the case of an event and working out scenarios for that. I think probably Mr. Wall could give you a lot more detail about that question, but it's obviously an issue and one that will be addressed in the hospital preparedness grant.

Mr. HORN. How about the Veterans Administration, are they involved in these committees and all the rest?

Dr. MILLER. They are.

Mr. HORN. How about it, General, are they at the table when you're coordinating things?

General WHITNEY. Sir, we don't really have an interface with the Veterans Administration other than our veterans affairs organization we have within the Department of Military and Veterans Affairs. Most of the interface that we have is the same thing that we have with public safety as well as public health and environment. So we go to the same meetings that they go to, but we don't have a direct interface with them.

Mr. HORN. When we started with the first of these series, it was in Nashville, Tennessee, and Vanderbilt Medical School and Hospital. And we had various diversions trying to get the rural parts of Tennessee involved as well as the urban. There is an incident where you had some strange thing that's attacked people, and they don't quite know what it is yet, and they want to bring it into an urban hospital and land the helicopter on the roof. We found out that even with all of the military forts and camps and you name it in Tennessee, the military helicopters could also sit on the roof. However there was no communication between them because of the difference frequencies.

And I'm wondering, General, have we got any feeling around that we've got some frequencies where people can go and involve the law enforcement very rapidly and so forth? How much of a problem is that?

General WHITNEY. Well, that is a significant problem. We have military frequencies, normally VHF and UHF radios and some FM radio capability in our military aircraft helicopters, as you talked to. We have significant Army aviation capability here in the State of Colorado within our Colorado Army National Guard.

However, we also have the capability to talk to other local responders through a digital trunk radio that has been issued to the Colorado National Guard, and we can give those to a helicopter that, say, has to land on the roof of a hospital somewhere if we need to. OEM has done a very good job, I think, of making sure that all the State emergency responders can communicate with each other in that respect.

Mr. HORN. Well, that's encouraging. I'm also interested in where we stand with laboratories other than CDC and other than the State of Colorado. Is there a use for the the various colleges' and universities' laboratories? Also the community colleges and the high schools if you're out in a rural situation, because Colorado is spread out and it's long miles to get to some of the things that we ordinary go and just think it's everywhere as in urban America or urban Colorado. How do we help the people in the rural part?

Dr. MILLER. That's a very good question, and the issue of surge capacity is also a huge issue. It doesn't take a lot for laboratories also to become overwhelmed if there's an event. And one of the things that we are doing within our bioterrorism grant is trying to bring all these laboratories together, the hospital laboratories, the CSU laboratories, the university laboratories. We've done a survey and tried to find out who has the capability, who needs to be trained in order to understand what tests you can do to rule out a bioterrorism agent, who needs to be trained about how to handle these things carefully because, obviously, you do. So that's a part of our work, to try and reach out to these other labs.

And we recognize the rural issue too, and that's why we are trying to increase the capacity in our local health departments, like Mesa Health Department, Weld Health Department, El Paso; even though that's a metro area, they serve a rural area. So we are trying to address that issue.

Mr. HORN. Now, CDC, as I believe, has a certification program in some of these laboratories. Is there a range of complexity? How does it work?

Dr. MILLER. Yes, there is a range. There are level A, B, and C laboratories. The State lab is a level C. That means we can quickly identify bioterrorism agents using advanced molecular methods. Level B laboratories have less advanced methods. Level A laboratories are basically clinical labs where they do hospital sorts of work. At that level lab, really the lowest level lab, they need to be able to rule out an agent, and say, "I identified it as X, therefore I know it's not anthrax." If they can't rule it out, they need to know how to package it and get it to us or get it to a level B laboratory where further work can be done.

Mr. HORN. Is there an overwhelming feeling yet in the Atlanta CDC where they're just overcome with people sending samples in and all this. And will that get done or will we have to do it in another way by using the States and the localities?

Dr. MILLER. I think during the anthrax event that was definitely the case. Laboratories were overrun. And people are definitely trying to figure out how to avoid that if we have another threat like that. And one of the things we need to do is work closely with the FBI, and we did that in Colorado. We need to make sure we have good threat assessment and that our laboratories are only used when there's a credible threat.

Some States didn't do that; they allowed every specimen under the sun into their labs, which is dangerous, first of all, because those specimens weren't screened well, but it also used up their capacities so that they couldn't respond if there had been a credible threat. So I think we actually did that pretty well in Colorado. We were not completely incapacitated during the anthrax event, and we want to continue to maintain that kind of policy and work closely with the FBI to do that.

Mr. HORN. Do you agree, gentlemen from FEMA and the FBI, on this?

Mr. CARBALLIDO. I agree wholeheartedly. The protocols that we set in place and worked very hard on were instrumental in the screening process, absolutely.

Mr. HORN. Has that been done around the Nation or is it just Colorado that's doing it?

Mr. CARBALLIDO. It has been done in many places but not everywhere.

Mr. HORN. Because of your protocols which you're very strong on, I come from Los Angeles County. We have had protocols, compacts, contracts, whatever you want to call it, for law enforcement, for fire enforcement, all of that, so we can help each other even if it's 500 miles north in the Santa Clara Valley or Central Valley and up to Stockton so where they need help. So do we have a lot of that here in Colorado?

You've got the big city here. And what about with the terrible things that have happened in a lot of these States, and one of them is Colorado, in terms of the fires and all, which puts tremendous pressures on trying to get something done. What do you hear on that and what do you think about it?

Mr. CARBALLIDO. We were not involved in the fire issue, sir. That, we were not involved in.

Mr. BAKERSKY. The protocols that we have in place for support from other Federal agencies, other State agencies. As far as the FEMA Region VIII, because of western-type climate, you know, large concentration in small areas of population, a lot of the protocols for mutual aid have been in place. A good example that we had, we actually tested most of our protocols during the Olympics in Utah. We brought in all of our resources, not just Federal assets, but also additional State assets, using the protocols—like Dr. Miller was mentioning the anthrax—that were actually used during some of the anthrax scares, during the Olympics.

So I think in this region we are fairly fortunate because of some of the activities that we've had. We've had the protocols in place between the law enforcement, between the emergency management agencies. Like with the Pope's visit, the G8 visit, the Olympics. So we've had a lot of real-world events that helped develop this partnership between Federal/State/local, not only in the law enforcement side of the house, but also in the emergency management/public safety side of the house. So they've been tested and they work fairly well. And a lot of our protocols are basically prototypes for some of the rest of the Nation, and also the activities that we did in the Olympics are now prototypes for other agencies throughout the United States.

Mr. HORN. Some people in other States have said there's just too much money being spent on planning rather than providing necessary equipment and training for those on the front lines of emergency response.

Can you give us a feeling of what's happening there? Is it just planning or are we getting the goods so people can do their job, be they a first, second, or third responder? FBI? FEMA? Anybody else want to take it?

Mr. CARBALLIDO. I would simply say that planning is also paramount. I don't know if there's a balance between equipment and planning in place, but planning cannot be underestimated. We exercise the plans, and I think that's why we were successful in responding to the anthrax threat, which was major, as you well know. So I really don't know if there's a balance that exists in Colorado between equipment, training, and planning, but planning is paramount.

Mr. BAKERSKY. One of the things we looking at with the new grant process is just not the planning, but the planning also includes equipment. When we are saying planning for equipment, we are looking as the interoperability of the equipment. We want to make sure that we have a standard. That's what we are trying to do with the new grant process is setting up a standard. Resource typing, so that if you request a certain thing from another area, you're getting the same. Some of the problems that you have in equipment, even on September 11th, things just as easy as hose thread. You had on Staten Island——

Mr. TANCREDO. As what?

Mr. BAKERSKY. Hose thread, thread on the end of a hose. The Staten Island Fire Department does not have hose that can be connected to New York City hydrants. So what they're trying to do— So you have the planning in equipment, but what you're trying to do is provide a standardization of equipment that could be used throughout the United States. A good example——

Mr. HORN. That's fascinating to me. These were borough differences of the five boroughs or so?

Mr. BAKERSKY. Yes, sir.

Mr. HORN. We do know Staten Island is different.

Mr. BAKERSKY. Yes, sir. I'm from New York originally, so——

Mr. HORN. And they didn't know about that until the problems came?

Mr. BAKERSKY. Probably not because they really did not—New York City being such a large organization, they didn't have mutual aid compacts in place. When you have a fire department of 18,000 individuals, they've never had an event that was beyond the scope of their capability, that they could not handle it with their own internal structures. September 11th came down and it did raise some issues.

There's other examples you could use, like scuba gear. You can go from one end of the country to another end of the country and the regulator on scuba gear is the same. You go to the fire fighting community and MSCA and 3M and SCBA equipment all might be different threads. So that's one of the things we are looking at besides the equipment, to have the planning to be sure that you have the interoperability of the equipment that can be used in response to an event.

Mr. HORN. Well, I'm glad you mentioned that because that's news to us. But you never would hear it in most places in America anyhow unless you've seen it there.

Any other little things like interoperability not happening? Or do you all have interoperability in Colorado? You're either sick of hearing it or is it done? And are there still gulfs somewhere that aren't getting treated?

Ms. MENCER. Mr. Chairman, I also serve on the national task force for interoperability as well.

Mr. HORN. Boy, am I going to follow you around.

Ms. MENCER. In my spare time I do that. And I'm happy to report, after listening to the other States that are represented on that task force, that we are far and away above many other States with our interoperability issues, and I think that's due directly probably to Columbine, which emphasized to us that we needed to be interoperable.

We have the digital trunk radio system in the State. We have been progressing through different sectors of the State to accomplish that. Unfortunately, our funding was stopped this year because of our fiscal problems again. But once we get our funding reestablished, we'll continue with our progressing across the State with getting our DTR capabilities up. But, as you know, it was just announced earlier this week, Senator Campbell effected this system that will be like a patch system for different radio systems so that those areas of the State that are not interoperable, they will be able to use this system to patch through and get them connected. So that is not a long-term solution to the problem of interoperability but is a short-term fix until we can get the digital trunk radio system up all over the State. So we are working very hard at that, and we have made great progress at that in the last couple of years.

Mr. CARBALLIDO. If I may add, Mr. Chairman, we also have a similar piece of equipment that we obtained—we were one of a number of offices in the FBI—from our research facility that accomplishes the same purpose, and this was done after Columbine as well.

Mr. HORN. Now, how much does this cost in terms of those that have equipment and need to be changed? And when the Federal Government gives a grant out, do you think they ought to say and demand it, that if you're going to use the taxpayers' money, it ought to be the right way, and figure out what are you going to do with the equipment that is not doing very well? How do you handle that?

Ms. MENCER. Mr. Chairman, this national task force is looking to make statements to bring forward to the Office of Homeland Security, to the President, stating what they believe are the best practices as far as interoperability goes.

I was amazed to learn when I attended my first meeting of this task force that some States don't even see the need for interoperability, which I think is amazing. But, of course, most States haven't had a Columbine incident to reinforce the necessity of this.

I think they will be coming forth with the statement encouraging that all States go to some level of interoperability, 800 megahertz, 700 megahertz, whatever it happens to be for that State. I think we will see all States coming on board with this eventually. But, again, I think Colorado is far ahead of that curve, and I'm happy that it is.

But it is a continuing issue. I think the Office of Homeland Security is looking at earmarking some funds particularly for interoperability, and that may be what we need as a State to continue with our progress with getting the other sectors up in line. So I'm hopeful that we'll see some funding in that regard.

Mr. HORN. Well, that's good because there's been some concern about the department created, that they haven't gone for standards against which one can then know we've done something right or we haven't. And so we are going to urge that a little bit and give it a nod.

There's a number of questions we have here that we might want to use for you. And let me just say, if you had 30 seconds with the President of the United States, went into his office, what would you say to President Bush is the most important thing on dealing with terrorism?

Let's just go down the line. General?

General WHITNEY. Well, sir, I guess as it relates to my specific mission area, I would ask him how we could implement new equipment, new training, and new parts of our organization in order to be able to meet the challenges of this terrorist threat.

Mr. HORN. How about it, Ms. Mencer?

Ms. MENCER. Well, I would first compliment him, I think, on what he's attempting to do with the Department of Homeland Security. I think focusing existing resources is what we need to do. And I think the resources are there, the capabilities are there. We need to have a place to focus those, and I think he's doing that, as we've tried to do in the State as well.

Mr. CARBALLIDO. Mr. Chairman, I think I would ask the President for a great deal of money and technology that exists presently to create data bases throughout the country that could better coordinate all the information that we all receive at the various levels, to better connect the dots and improve on our intelligence base, because for us that is the key to prevention and to be in a proactive posture so that we don't have to involve ourselves in crisis management.

Mr. HORN. Mr. Bakersky.

Mr. BAKERSKY. I would stress that we continue the initiatives that were started with the fiscal year 2002, supplemental fiscal year 2003, providing resources, both monetary and personnel resources, to States and locals, which basically are the first responders. They're the individuals that are going to be putting their—everything on the line. When we have an incident, we have to make sure that we have funding streams in the initiatives that are started and maintained for the next 3, 4, 5, 6, however many years it takes.

Mr. HORN. Dr. Miller.

Dr. MILLER. I would panic, first of all, but——

Mr. HORN. No, you wouldn't. He's a really friendly guy.

Dr. MILLER. He does seem like a really friendly guy.

I would echo some of those comments, and I would stress that this is really a new role for public health. And if we are really going to develop this capacity in public health to respond to bioterrorism and to be part of emergency plans, we need long-term infrastructure support.

And I would also try and frame public health as part of the first responder community. I think it's easier to understand that way that we also need to be prepared to be first responders, and that's a new role.

Mr. HORN. Very good. Any more thoughts?

Mr. TANCREDO. Nothing more. Thank you, Mr. Chairman. Very, very interesting.

Mr. HORN. And we have with us the General Accounting Office, and at the end of the next panel. We bring them here because we've got over 50 blue books already, and it's very worthwhile material, if you don't have it, and I would hope GAO would send it to all of you. And we ask them, what haven't we done? Where are the openings that we don't know what we are talking about? And then go back to it. So that will come up after the next panel.

So you've done a wonderful job, all of you. And I think Colorado seems to be in good hands. So we will now move to panel 2.

[Recess.]

Mr. HORN. Recess is over. And before we begin with Panel 2, there is a statement, a very fine statement, by Representative Mark Udall, and I would like the reporter to put that following the Horn and Tancredo statements at the beginning of the thing. We'll put Mr. Udall's in as it is, and she will give it to you.

I just want to make sure everybody is here. We've got Mr. Wall, Mr. Sullivan, Lieutenant Hoffner, Lieutenant Wicks, and Mr. Posner, so you know this procedure. Since it is an investigative committee, if you have any staff to support you, have them take the oath so I don't have to do it in the middle of the areas. So if you'll raise your right hand.

[Witnesses sworn.]

Mr. HORN. The five witnesses have affirmed the oath, and we will begin with Mr. Wall. If some of you haven't been here in the first session, we were going right down the line in order. And as we call your name, your whole, full written statement is automatically put in to the report, so you don't have to give every word in it, but we would like to have you give us 5 minutes or so, or maybe 10 sometimes, if it's that wonderful, and we would like you to give us the summary of it, and then we can get into the question period at the end, and we'll do that in each case. We've looked at the documents; they've been very good.

And so we'll start right now then with Larry H. Wall, president of the Colorado Health and Hospital Association.

STATEMENT OF LARRY H. WALL, PRESIDENT, COLORADO HEALTH AND HOSPITAL ASSOCIATION

Mr. WALL. Mr. Chairman, thank you very much for the opportunity to be here. In addition to my responsibilities as president of a hospital association, I'm also a member of the Governor's Epidemic Emergency Response Committee, and I chair the Hospital Preparedness Advisory Committee. I appreciate the opportunity to testify before you this morning.

September 11th has obviously created a new world for all of us, and that includes hospitals. Hospitals as first-line responders have always been the foundation of a response to the medical needs of patients, be they personal or as a result of natural disasters or flu epidemics.

Historically, the resources of the hospital system have been adequate to meet the needs. The potential use of weapons of mass destruction and bioterror agents, however, results in the need for a whole new level of preparedness. There are at least eight areas that need to be addressed. One, communication and notification, and we've heard a lot about that already this morning. Communication for hospitals is as critical as it is for other organizations and agencies. Disease surveillance and reporting and laboratory identification. That was referred to earlier as part of the intelligence network, which I think is an important issue. Personal protective equipment. Facility enhancements. Decontamination facilities. Medical, surgical, and pharmaceutical supplies. Training and drills. And mental health resources. At this particular point the resources are not adequate to address all of these needs.

Just to use a very simple example, the 2-year HRSA allocation for Colorado is approximately $4.5 million, or roughly $70,000 per hospital. The estimated cost to address the communication issue alone is in the neighborhood of $3.5 to $3.7 million, leaving little for the remaining seven areas of need.

The current allocation of dollars, while it is very much appreciated, is really inadequate to meet the needs with regard to hospital preparedness. I think it's important to understand, however, that the public can certainly be assured that hospitals will be as prepared as they possibly can within the constraints of the available resources. Neither Congress nor the American public should assume that at the current level of Federal funding that hospitals will be fully prepared to handle the outcome of a significant event. We are significantly ahead of where we were on September 11th, and progress on preparedness will continue to be made. But more funding is needed if hospitals are to meet what I believe are the preparedness expectations of Congress and the American public.

I want to thank you for the opportunity to testify, and I'll certainly be happy to address specific questions with regard to hospital issues, some of which were raised in the earlier testimony.

[The prepared statement of Mr. Wall follows:]

August 18,2002

Members of the Committee on Government Efficiency, Financial Management and Intergovernmental Relations

I am Larry Wall. I am the current President of the Colorado Health and Hospital Association. I am also a member of the Governor's Expert Epidemic Emergency Response Committee and Chairman of the Hospital Preparedness Advisory Committee. I appreciate the opportunity to meet with you and share my thoughts and observations about hospital preparedness with you.

September 11, 2001 has created a new world for all of us not only in our personal lives, but our professional and community lives as well. All of us view our security very differently than we did one year ago and the same is true for hospitals.

Hospitals have always been the foundation of a response to the medical needs of patients brought about as a result of personal injury, injuries caused by natural disasters, and other medical conditions such as those resulting from flu. Historically the resources available in the hospital system have been adequate to address these need. The threat of the use of weapons of mass destruction and bioterror agents results in a whole new level of preparedness for hospitals not historically present. Since 9/11 hospitals are actively engaged in preparing for such an event, however, the resources available to fully prepare for such an event are scarce. As a result of significant reductions in acute beds over the last 10 years, current workforce shortages, and a lack of funding for the purchase of equipment and training our ability to respond to a significant event is limited. Preparing for a response requires eight key areas to be addressed by hospitals. Each area has its own challenges and needed resources. The areas are :1) communication and notification; 2) disease surveillance, disease reporting and laboratory identification; 3) personal protective equipment; 4) facility enhancements; 5) dedicated decontamination facilities; 6) medical/surgical and pharmaceutical supplies; 7) training and drills; and 8) mental health resources. The American public can be assured that if such an event occurs, hospitals will be there; will do everything humanly possible to respond; and will do so to the maximum extent resources will allow.

First responders and hospitals are the first line of defense in case of an event. When one considers the potential variety of ways in which such an event could be initiated, the preparedness of hospitals becomes even more complicated. Given the potential for the use of nuclear, chemical and biological agent's hospitals need to be prepared for an "all hazards" response. The use of personal protective equipment, decontamination equipment for all hazards, isolation rooms in both the ED and on inpatient units, reverse air flow rooms, the expansion of emergency power, communication requirements, etc., result in the need for significant capital investment. While all hospitals need not be prepared to handle the full impact of such an event, each hospital must be at least minimally equipped to respond on a local basis until support can be provided to them from a regional and

EXHIBIT
9
HUNTER & GEIST

state basis or prepared to support and assist other hospitals. It is also important to understand that doing what is necessary to respond will likely take several years to accomplish given the need for construction, extensive training, purchase of equipment, etc. Another complicating factor is that as preparedness modifications are done to facilities they will be required to meet the most current life safety codes (many of our hospitals were constructed in the 1960's and early 1970's), which will increase the cost of construction.

While the current allocation of dollars for hospitals is a clear recognition of our important role in preparedness and response and is appreciated, the funding is woefully inadequate to meet the needs. If we assume that all hospitals in Colorado have needs (a fairly strong assumption), our first year funding would result in approximately $24,000 per hospital if evenly distributed. If we hold the first year funding and add to it the second year funding (which we are planning on doing), we will have approximately $70,000 per hospital for the two years. More but nowhere near what is necessary. The American Hospital Association in an assessment of its membership estimated the need to be $11 billion nationwide. If allocated evenly among the states, Colorado's needs would approach $200 million in total. If we look at only **one** of the eight areas listed above, communications (a very key area), the estimated cost is $37,500 for a non-metropolitan hospital and $75,000 for a metropolitan hospital. In total the estimated cost for preparedness for all eight areas for a metropolitan hospital is $3 million and for a non-metropolitan hospital is $1.4 million. As you can see there is a significant funding gap.

Colorado and America's hospitals will be as prepared as possible within the constraints of the resources available to do so. It is important to understand that the current level of funding is a beginning but a lot more is needed. The American public should not assume that hospitals, as a result of current Federal funding, are fully prepared to handle the outcome of a significant event. We are significantly ahead of where we were on 9/11 and we will continue to make progress on preparedness, but more funding is necessary if we are to meet what I believe to be the expectations of Congress and the American public.

Thank you for your time and attention and I will attempt to address any questions you may have.

testimony

Mr. HORN. Thank you very much, Mr. Wall. We now have David B. Sullivan, acting director, Office of Emergency Management for the city of Denver.

STATEMENT OF DAVID B. SULLIVAN, ACTING DIRECTOR, OFFICE OF EMERGENCY MANAGEMENT, CITY OF DENVER

Mr. SULLIVAN. Good morning. Thank you for the opportunity to speak to you today. The events of September 11, 2001 have greatly exacerbated the threat to this country from terrorist attacks. Congress's willingness to provide leadership, direction, and funding to support our preparedness efforts has been greatly appreciated. We at the local level are thankful for the support we've received from Congress and the administration. Our efforts in Denver began prior to September 11th with the Nunn-Lugar-Dominici legislation. The equipment, training, and support we have received through that program has greatly enhanced our preparedness; however, there is still a great deal of work to be done.

The $3.5 billion allocated for first responders in the President's homeland security strategy is truly needed, and we applaud the efforts of the administration and Congress, but there are some concerns. First and foremost is the competitive nature of the grant process that pits local first responders against each other for the Federal funds. This has been a problem in the past with the Department of Justice grants and continues to be troublesome.

Terrorism events will tax the full resources of local jurisdictions, States, and the Federal Government. We must be prepared to respond in a comprehensive manner utilizing all resources available. The burden for response lies squarely on local jurisdictions. State and local Federal resources are sometimes hours or days away. The capability of the initial response is what will save lives. Rather than fund specific first responders, funding should be made available through local jurisdictions to provide for all their needs rather than preidentified disciplines. The breakdown of the funding in the homeland security strategy identifies how the moneys will be distributed. The distribution is similar to past DOJ programs in that only 9 percent of the $37 billion allocated for homeland defense will go to local jurisdictions. Of the moneys going to the States for pass-through funding, 25 percent will remain at the State level for whatever requirements the States determine. However, the 75 percent passed through to local jurisdictions are predetermined, fitting into defined categories of planning, equipment, training, and exercises. Amounts are predetermined for each category and the State determines priorities. Unfortunately, each jurisdiction's different, with different threat levels, different levels of preparation, and different resource needs. As such they should be given the same opportunity the States have in determining how the moneys are utilized.

Emergency management, by its very nature, must integrate and collaborate with all the players involved on issues of domestic preparedness. Funding is required not only to provide initial resources, but also to create an infrastructure of domestic preparedness that will be a long-term integrated component of the day-to-day operations of an emergency management system. Federal support for local emergency management programs have slowly eroded over the past 15 years to the point where many emergency management programs have either been dissolved or incorporated into other supplemental responsibilities of another municipal agency or department. Local emergency management programs have always been the forgotten stepchild of other public safety agencies in terms of funding and authority. While the various and numerous Federal agencies have provided grants to traditional first responders, there has not been funding directed to support local emergency management agencies. Federal funds to build, maintain, or improve local emergency operation centers ended nearly 10 years ago, and there are little or no local funds to pay for capital improvements to such centers. If local communities are truly expected to be the first line of protection in the new homeland defense system, they must be properly equipped to facilitate efficient and effective decision-making in an adequate emergency operations center. Funds are needed to immediately—funds are needed immediately to upgrade and establish a full functional EOC in each community.

While we at the local level support the enhancement of homeland security on a national level, we must never lose sight of the normal day-to-day emergencies and disasters that affect and could potentially devastate our communities. Attention and resources for floods, tornadoes, winter storms, wildfires, and other natural technological hazards must not be diminished at the expense of today's hot topic of weapons after mass destruction and homeland security.

As local emergency managers, we stand ready to assist by coordinating the planning, training, and resources of our jurisdictions in developing comprehensive readiness programs. We look to you for assistance in developing these programs, recognizing the critical coordination role that local and emergency managers play in developing readiness programs. We do not operate as single disciplines when preparing for or responding to natural disasters, and we certainly should not do so within the arena of terrorism.

Again, I thank you for the opportunity to speak to you today.

[The prepared statement of Mr. Sullivan follows:]

TESTIMONY FOR HOUSE COMMITTEE OF GOVERNMENT REFORM'S SUBCOMMITTEE ON GOVERNMENT EFFICIENCY, FINANCIAL MANAGEMENT AND INTERGOVERNMANETAL RELATIONS

DAVID B SULLIVAN, CEM
ACTING DIRECTOR
MAYOR'S OFFICE OF EMERGENCY MANAGEMENT
CITY AND COUNTY OF DENVER

AUGUST 23, 2002

The Events of September 11, 2001 have greatly exacerbated the threat to this country from terrorist attacks. Congress' willingness to provide leadership, direction and funding to support our preparedness efforts has been greatly appreciated. We on the local level are thankful for the support we have received from Congress and the Administration. Our efforts began prior to September 11th with the Nunn-Lugar legislation. The equipment training and support we have received through that program has greatly enhanced our preparedness. However, there is still a great deal of work to be done.

Let me first provide some background on emergency management programs. Local emergency management directors are responsible for the development of comprehensive emergency management programs in counties, cities and special districts throughout the United States. This involves the development of mitigation, preparedness, response and recovery programs for every community in the United States.

Emergency Management programs at the local level are responsible for providing overall predisaster planning and mitigation programs such as training and exercising for natural and manmade disasters that potentially can affect a community. The Emergency Management profession as a career is relatively new in the United States history however, the tenets upon which the profession is based can be traced back to the Cold War days. While most local jurisdictions developed their first emergency management programs during the Cold War under the name Civil Defense or Civil Preparedness, the basic concepts of civil protection remains consistent. However, what has not remained consistent is the level of federal support to local jurisdictions to develop and sustain programs.

The $3.5 billion allocated for first responders is truly needed and we applaud the efforts of Congress but there are some concerns. First and foremost is the competitive nature of the grant process that pits local first responders against each other for federal funds. This has been a problem in the past with the Department of Justice Grants and continues to cause concern.

Annunciating the role of Emergency Management in the grant process could eliminate this competitiveness. Local emergency managers, as stated earlier, are responsible for the development of comprehensive, all hazard emergency management programs in their jurisdictions. This involved the development of mitigation, preparedness, response and recovery plans, including consideration of terrorist issues. Emergency management represents the linkage between first responders such as fire, police and EMS, to other parties involved in disasters, including local agencies, state and federal resources. In addition, local emergency managers are responsible for the comprehensive capabilities of their jurisdiction. It is their responsibility to ensure that all agencies are prepared equally to provide a balanced, unified response, as well as nontraditional responders such as public health, public works and human services.

In addition, there is a tendency to lock funding into specific disciplines and other jurisdictional needs are ignored. Terrorism events will tax the full resources of local jurisdictions, states and the federal government. We must be prepared to respond in a comprehensive manner utilizing all resources available. The burden for response lies squarely on local jurisdictions. State and federal resources are sometimes hours or days away. The capability of the initial response is what will save lives. Rather than fund specific first responders, funding should be made available to local jurisdictions to provide for all their needs rather than pre-identified disciplines. The break down of the funding in the Homeland Security Strategy identifies how the monies will be distributed. This distribution is similar to past DOJ programs in that only 9% of the 37 billion allocated for homeland defense will go to local jurisdictions. Of the monies going to the states for pass through funding 25% will remain at the state level for whatever requirements the state determines. However the 75% passed through to local jurisdictions are predetermined, fitting into defined categories of planning, equipment, training and exercises. Amounts are predetermined for each category and the State determines priorities. Unfortunately, each jurisdiction is different with different threat levels, different levels of preparation and different resource needs. As such they should be given the same opportunity the states have in determining how the monies are utilized. We also need to include the CDC grant process in our comprehensive efforts. The Center for Disease Control process, working independently of other federal programs promotes the development of separate health and emergency management systems. Working independently will ensure multiple, uncoordinated systems and responses. We need to include public health and hospitals in comprehensive planning efforts. Creating separate systems through the grant programs will only create duplicative stove pipe systems.

We also need to include Information Technology and Cyber Terrorism planning in the process, as well as incentives to develop continuity of government recovery and contingency plans.

Sine the Nunn-Lugar-Dominici program of the mid 1990's local emergency management agencies have been either the direct or indirect recipients of federal resources to help build the infrastructure to deal with the Weapons of Mass Destruction threat. Local directors have done yeoman work to ensure the needs of first responders have been and are being met. They have served as referees between first responders agencies in prioritizing equipment requests and as facilitators to bring public health and hospitals into a comprehensive system. Denver's efforts in developing a functioning five county Metro Medical Response System is a prime example. The Metro Medical Response System originally funded by the United States Public Health Services as part of Nunn-Lugar, has united the Denver Metro area in a comprehensive response system. Future funding for this system is unclear. We strongly support sustaining these funds for the MMRS program.

Emergency Management, by its very nature, must integrate and collaborate with all the players involved on issues of domestic preparedness. Funding is required not only to provide initial resources, but also to create an infrastructure of domestic preparedness that will be a long-term integrated component of the day-to-day operations of an emergency management system. Federal support for local emergency management programs has slowly eroded over the past fifteen years to the point where many emergency management programs have either be dissolved or incorporated into other supplemental responsibilities of another municipal agency or department. Critical to effective operation is dependable funding to support training, plans, facilities, staff, equipment and a communications net work. At best, the national systems is only adequately funded, while the local emergency management infrastructure is woefully under funded.

Local emergency management programs have always been the "forgotten stepchild" of other public safety agencies in terms of funding and authority. While the various and numerous federal agencies have provided grants to traditional "first responders" such as fire/EMS/police, public health and utilities, there has not been funding directed to support local emergency management agencies. Federal funds to build, maintain or improve local emergency operations centers ended nearly ten years ago and there are little or no local funds to pay for capitol improvements to such centers. If local communities are truly expected to be the first line of protection in the new Homeland Defense system, they must be properly equipped to facilitate efficient and effective decision making in an adequate emergency operations center. Funds are needed immediately to upgrade and establish a full functional EOC in every community.

While we at the local level support the enhancement of Homeland Security on a national level, we must never lose sight of the normal day-to-day emergencies and disasters that affect and can potentially devastate our communities. Attention and resources for floods, tornadoes, winter storms, wildfires and other natural and technological hazards must not be diminished at the expense of today's hot topic of Weapons of Mass Destruction and Homeland Security.

We welcome a more active and direct federal-local partnership, but we need to continually emphasize that the partnership must be provided adequate resources to be effective. Homeland Security is nothing new to most local emergency managers. We have been doing many of these efforts for many years.

We truly appreciate the efforts of Congress in taking the lead in preparing the nation for potential acts of terrorism. As local emergency managers, we stand ready to assist by coordinating the planning, training and resources of our jurisdictions in developing comprehensive readiness programs. We look to you for assistance in developing these programs. Recognizing the critical coordination role that local emergency managers play in developing readiness programs would be helpful in these efforts. We do not operate as single disciplines when preparing for or responding to natural disasters and we certainly should not do so within the arena of terrorism.

Again, thank you for all your efforts. Please feel free to contact this office if have an questions or concerns.

Mr. HORN. Thank you, Mr. Sullivan. Lieutenant Roger E. Hoffner, Arapahoe County officer of emergency management. Glad to have you.

STATEMENT OF LIEUTENANT ROGER E. HOFFNER, ARAPAHOE COUNTY OFFICER OF EMERGENCY MANAGEMENT

Mr. HOFFNER. Good morning, Mr. Chairman. Thank you. I apologize I don't have a prepared presentation. I was under the understanding that my sheriff was going to be presenting this morning, so I bowed to him.

What I'd like to do is—and what Dave Sullivan said, I agree to wholeheartedly. The emergency managers in the metropolitan area work very closely together, and what he said is right on track. I'd like to say a little bit about where I came from before I get to where we are and where we are going.

In 1996 I became the emergency manager for the county. Before that I was a deputy for over 15 years. And up to that point the only thing I knew and was aware of is what the responsibilities were for our response on the street to those everyday calls. I had no understanding or very little about incident command. I had very little understanding about preparedness for big disasters, none of that. And when I took the emergency management position, I found out that the job was probably more massive than I ever imagined, and it's more massive than I still imagine. I have never had a job that I haven't been able to master in 6 months to a year, and I've been doing this for 6½ years, and there are days I still feel lost. It's an incredible job.

In 1998 I had an opportunity to go to an exercise sponsored by the Department of Energy in Las Vegas. It was then that I started learning about PPE, personal protective equipment. They talked about the Quick Masks that every capital police officer in D.C. was wearing on their belt. I took that back and decided to do a grant to try to get gas masks. Because if we look in most of our police cars, they'd be lucky if they had an old military gas mask that's a false sense of security; they have nothing.

So with that money I applied for, $45,000, I got 400 gas masks. And the other thing that went with that is I tried to find out from government what was the best thing I could get for my money. And they all said, "Well, now, we could give you a list of things, but we can't tell you which one because that's a conflict of interest." Well, I found that very frustrating because I wanted to spend the best money that I could, get the most I could out of it, and be able to protect my people. So I did my own research and I bought 400 masks with the best chem/bio filter they had and distributed it to our people in Arapahoe County, but I was still 1,000 gas masks short of what I needed. When we started——

Mr. HORN. What do you have now?

Mr. HOFFNER. Right now I have about 850 bags out there, and that's where I'm going to go to. One of the things that we started with, where we really picked up our information, is the federally funded Top-Off 2000 exercise that came to Denver with the biological release. I was involved in that from the beginning with the planning phase, up through the incident command, and with the hot wash at the end to do a critique on.

A lot of people thought that was a failure. I thought it was an incredible success because what we did was we learned and we learned and we learned some more. And one of the big things we learned was that we don't have a good communication system. And we never worked that closely with the department of health, with hospitals, and we have no communications with them. We learned an awful lot from that and we are building on it. And when Aurora had the Nunn-Lugar money come down to do their exercises, one of the last ones they did about 4 months ago was almost identical to Top-Off, and the responses we had to that showed me that it's working, that people were working together, that we were talking about communications, that we were talking about our response, we were talking about mass distribution of medical aid to people.

So it's working. That money was very well spent.

The MMRS, the Metropolitan Medical Response System, was another grant that came down, and Aurora and Denver both were cities of that. And with Aurora getting the last one, we've had incredible response about making sure that money went to good use. We put stockpiles of supplies, Mark 1s and medical supplies for first responders to an incident. A major success, I think, for our front range.

And that other money that we used is—I had an opportunity to take our county back to Emmitsburg, Maryland, to the Emergency Managers Institute for an integrated emergency management exercise in the November 2000. They were all leaders of our county, all taken care of by Federal funds. That brought us together. There were people there I didn't think they could be in the same room, and as a matter of fact, when it came down to it, they worked very well together. So I've had the opportunity to meet with that group and followup every 3 to 6 months with followup meetings, with training, with tabletops, with that kind of thing, and that has been incredible. All federally funded money. Again, a success.

But then I get to the money that came down for 1999, 2000, 2001, and my image, when I think about New York City, is, sure, I see the plane going into the building, but my biggest image is seeing those cops and firemen with handkerchiefs tied around their faces and gagging and coughing and can't breathe.

So what I did with that money when it came down, Arapahoe and Douglas Counties did a regional application. I got—$352,000 is what was targeted for us. And I had in my mind that I wanted to put together some kind of first responder bag for these first responders, for the cops that are going to be out there first. We needed to take care of them.

When September 11th happened, my money was sitting someplace between here and who knows where. So I asked the Office of Emergency Management, the State, if they would request $113,000 to be expedited so I could get those bags together as quickly as possible. And they did that. And with that 113,000, I put 805 bags together, which included a chemical/bio mask, a chem suit, goggles, three different kinds of gloves, a decontamination kit, eyewash, earplugs, disposable mask, and put it all in a canvas bag that they could carry in their car. I assigned one of those to 805 cops. Still short, but I picked the primary, most probable officers that would respond and said, how many do we need? And I gave them out. They were very well received.

The only problem that comes up is we have a question about training. How do we meet OSHA standards? Well, reaching out to my resources that I have met over the last 6 years and my contacts, I had a meeting 2 days ago with Tri-County Health and with three representatives from National Jewish Hospital. They came up with a plan to implement training to include medical, limitations, and fit tests, which are the three primary things for OSHA.

They're going to do a proposal to me. They're going to do it on a very reasonable fee, and I'm going to include that in the grant, and it's going to cover every person that received a mask. And we are going to be able to take those and be able not to have—and one of the reasons why I went with this is, it's multifaceted; we can use them for the WMD event. But we can also use them for that everyday thing that happens, the turned-over tanker, with a meth lab, or whatever that they're doing perimeters on, to protect themselves. Some protection. They have nothing.

The next round of money that's coming now, we are hoping to get somewhere around $500,000, which I'm going to try to get probably another 1,000 bags put together to cover every cop in two counties. And now I've added one of the rural counties, Elbert County, into it because they pretty much have nothing. And I thought, well, being big brother, we can help them out. So we are going to bring them in, and I'm going to make sure every cop gets it.

The extra money that we've had with that—we call it extra because it's just in addition to that—is that our HAZMAT teams and our fire departments have been reinforced with detection equipment, with decon equipment, and personal protection equipment. And although we have not met every need that we would like to have, we have far exceeded—and I can't even imagine how far we've exceeded—over 1996 when I started.

So, yeah, we could use more money. We could use more money all the time, and there's other things we could do. But with the money we've had, we've come a long way.

We talked about interoperability with radio systems; that was one of the comments that was made earlier, the interoperability of communication systems. What they do is they plug in a radio and it allows them to talk to each other. But if there's no control on who gets on those radios, there's going to be such mass chaos that you won't be able to communicate, but you've got one frequency now or maybe two.

The other question we need to deal with is the incident command system, to have that somehow supported by Federal Government down to the State level that says that, yeah, we strongly suggest that everyone uses the incident command system, which includes a communications plan within that system. So that we know that when you have this one frequency, the only person talking on it is a commander, and everybody else talks with other frequencies within their own department. And we can do that. But I think if we don't have that communication, we are going to be lost, just like we were with Columbine.

And on the grants, real quickly, I agree with Dave that when these grants come in, it's hard to look at a grant when we say we need communications and the grant is telling me that we need to spend $500,000 on needs assessment. We know what the needs are; we don't need to have somebody tell us that we need to do a needs assessment.

Mr. HORN. This is—which agency made that statement?

Mr. HOFFNER. Which one?

Mr. HORN. On the one that you would have to do a needs assessment.

Mr. HOFFNER. That comes down on a lot of grants. They have areas that say specifically you have to spend this much money on, and we are saying, "No, we'd like to spend it on this." And they say, "No, our guidelines are there." And what I'd like to see is be able to—like Mr. Sullivan said, is to be able to have those grants so they're a little more open. And the new one coming down has a little more latitude to it. And it's going to allow us to be able to do a little more approach to exact needs of our local agencies, and I think that's critical.

And I'll shut up except for questions. Thank you.

Mr. HORN. Very interesting.

[The information referred to follows:]

July 31, 2002 | Vol 1. Issue 13

Homeland Defense Journal

"He is best secure from dangers who is on his guard even when he seems safe." —Syrus Publilius

Homeland Defense Journal, Inc. | Suite 1003 | 4301 Wilson Boulevard | Arlington, Virginia 22203
www.homelanddefensejournal.com | Phone: 703-807-2758 | Fax: 703-807-2758

WHAT'S INSIDE

OUR STAFF

PUBLISHER
Don Dickson
ddickson@homelanddefensejournal.com
301-455-5633

EDITOR
Marianne Dunn
mdunn@homelanddefensejournal.com
703-807-2495

CIRCULATION
David Dickson
dicksond@homelanddefensejournal.com
703-807-2758

REPORTING STAFF
Steve Kingsley
skingsley@homelanddefensejournal.com
703-807-2758
Kelly Kingsley
kkingsley@homelanddefensejournal.com
703-807-2758
George Groesbeck
ggroesbeck@marketaccess.com
941-360-3663
Tony Rahimi
trahimi@homelanddefensejournal.com
703-807-2758

GRAPHIC DESIGN
Dawn Woelfle
dwnmrie@graphic-designer.com
941-746-4923

ADVERTISING AND SPONSOR SALES
Cara Lombardi
clombardi@homelanddefensejournal.com
703-807-2743

First Responders: A self-examination

By Jerry A. Lema
For Homeland Defense Journal

The mission of the Naval War College, Newport R.I., includes contributing to the national security of the United States by supporting the U.S. Navy and U.S. defense community with research, analysis and war gaming. In the past, the U.S. defense community would be defined as the armed services, the Office of the Secretary of Defense, and the intelligence agencies. However, since the terrorist attack last September, domestic first responder organizations, such as police and fire departments, must be considered part of the U.S. defense community. For that reason, the War College, and specifically its war gaming department, has become active in supporting the domestic organizations (municipal, state and federal) charged with responding to terrorist attacks on the U.S. homeland.

The war gaming department has designed and implemented a number of war games for first responders from the New York City Fire and Police Departments and the Rhode Island Emergency Management Agency. The objectives were to assess command and control implementations and decision making and gaming exercises conduct critical self-examinations of those agencies' organizational and operational preparedness to deal with future attacks. Similar exercises are scheduled for New York's Office of Emergency Management and the Federal Emergency Management Agency Region 1.

The war gaming department applied its experience in scenario design, event facilitation and analysis in providing the agencies with a forum to evaluate the development of contingency plans and the subsequent need to employ them in reaction to a significant threat. These simulated environments offered opportunities for both leadership and staff to gain insight into the strengths and weaknesses of their organizations.

In the February 15, 2002, issue of *Navy Newstand*, Dr. Kenneth Watman, director of the department, said, "We discovered [as a result of the FDNY exercise] there are a lot of similarities in how

continued on page 3

Mobile Solutions Enhance Homeland Defense Initiatives

By Richard Owen
For Homeland Defense Journal

The success of the nation's homeland defense initiatives will rely heavily on the ability to rapidly move information, materials and individuals efficiently and effectively around the country and the world. Within our own borders, we have checkpoints and processes to protect and a need to ensure that appropriate items accounted for as they move by air, rail, sea and ground.

For example, recently a heavily guarded shipment of plutonium on its way through South Carolina became part of a national debate over transporting nuclear material in the new age of terrorism. Questions ranged from: "Does anyone know how much is being transported?" to "Is any of it missing?" and "Did it make it there safely?"

Public sector IT strategies reflect a growing need to improve business processes throughout – and across – government agencies. By automating and mobilizing processes that were previously paper based and labor intensive, organizations could achieve new efficiencies and streamline cumbersome procedures.

According to Bob Day, U.S. Coast Guard West Coast CIO, "The current challenge is how to take all of our paper-based processes and automate them agencywide, so everyone is more efficient."

Today, the U.S. Coast Guard has between 1,000 and 2,000 inspections happening concurrently. These inspections include boarding incoming vessels, which requires specific checklists to be completed and filed. In the past, all of this has been a lengthy error-prone paper process.

continued on page 7

Planning for Emergencies in Colorado

By Lt. Roger Hoffner
For Homeland Defense Journal

Long before Sept. 11, the Colorado Office of Emergency Management and local governments planned and prepared for a potential terrorist attack. Arapahoe County's Office of Emergency Management (ACOEM) has been an integral part of this process.

Nearly two years ago, ACOEM learned of grant money that would be available for counties to direct toward their first responders: police, fire departments and HAZMAT teams. The first step was to complete a countywide assessment listing targets, vulnerabilities, capabilities and needs. ACOEM and nearby Douglas County started the work on a regional assessment and application. In August 2001, Arapahoe/Douglas Counties were awarded $352,000, but the money would take a few months to filter from the Department of Justice to the state and eventually to the county.

The grant fell about $550,000 short of the listed assessment needs, however, the money was viewed as a great start to prepare.

As the county's emergency management coordinator, I attended many classes on weapons of mass destruction and terrorism. My plan was to use part of the grant to put together a "first responders bag," containing a chemical/biological gas mask, three types of gloves, goggles, disposable respirators, eye wash, ear plugs, chemical suit with hood, overshoe booties and duct tape. After the terrorist attacks, on behalf of ACOEM, the Colorado Office of Emergency Management asked the DOJ to expedite $113,000 for the bags. After approved, 805 bags were put together and distributed to all the law enforcement agencies in Arapahoe and Douglas Counties. Since then, ACOEM has given 32 additional bags to Colorado State Patrol at the Castle Rock Office.

In addition to the grants, the federal Office of Emergency Management coordinated an integrated emergency management course, county specific, at the Emergency Management Institute in Emmittsburg, Md. In November 2000, 79 participants — mostly from Arapahoe County, which included three county commissioners — experienced the realistic exercise and training that affected their own county. Since then, the group has committed to continue the training. In May 2001, the group convened to discuss what it had learned and where it wanted to go.

In October, the group again gathered to discuss how to handle recovery if impacted by a large tornado and conducted a tabletop exercise. In March, the group met for a presentation from Greg Moser, Colorado Office of Emergency Management counterterrorism expert, about weapons of mass destruction and terrorism, as well as an update on how Colorado is preparing. Again in May, the group gathered to discuss its plans for a terrorist attack and test plans and call down lists.

Those of us in the first responder community identified two things that seemed to fall apart on any large, multiagency incident: communications and incident management. We have seen this at recent events, such as the high school shootings in Columbine and the manhunt for a "cop killer" in the Cortez area of

continued on page 20

Planning for Emergencies in Colorado

continued from page 19

Colorado. The Incident Command System (ICS) helped manage these two problems.

The ICS is modular in structure and could be used for large or small incidents, emergency or non-emergency. The system is built on a communications plan, a unified command structure if more than one agency is involved, and resource and personnel management. First responders must understand ICS, then train and exercise with it. In Arapahoe County, all the sheriff's office deputies are trained and the other police agencies are either trained or being trained.

Is Arapahoe County prepared for a terrorist attack? Maybe not, but it is far beyond many in most of the country. ACOEM — and 56 of the 64 counties in Colorado — established a terrorist annex for their county emergency operations plan. And all of the public health agencies in Colorado have completed assessments relating to health capabilities and needs.

The key is to realize that although terrorism is on everyone's minds, first responders plan for disaster everyday.

Lt. Roger Hoffner is Arapahoe County Emergency Management Coordinator

DoD 'OK' in Radio Bandwidth Transfer to Private Sector

By Gerry J. Gilmore
American Forces Press Service

The Department of Defense retains access to valuable radio bandwidth needed for national security although the government gave up a segment July 23 to facilitate growth in the U.S. telecommunications industry.

The Department of Commerce announced its plan July 23 called the "3G (3rd Generation) Viability Assessment." DoD and some other government agencies will transfer 45 megahertz of radio bandwidth to the private sector. The frequencies will come from the 1710-1755 MHz range.

One of the challenges in developing the 3G plan was how to reallocate bandwidth without impairing DoD's network-centric warfare and information superiority missions, according to Commerce Department officials. Military transformation calls for quantum leaps in the use of computerized information technology that depend on wireless systems.

However, the bandwidth transfer won't hurt DoD's missions, said Steven Price, deputy assistant secretary of defense for spectrum, space, sensors and C3 (command, control and communications) policy.

"We welcome the findings in the 3G Viability Plan and believe the plan supports the needs of national security," Price noted. He added that DoD would be reimbursed for associated costs in transferring the bandwidth to gaining private-sector entities.

Price noted the plan "requires some changes" to certain military systems, but said DoD doesn't lose because it will have access to more bandwidth, if needed.

"DoD believes that implementing the 3G plan will (neither) degrade military capabilities nor harm national security interests," he remarked.

The plan acknowledges "defense's growing spectrum needs, and we expect [the National Telecommunications and Information Administration and the Federal Communications Commission] will continue to take these needs into account in the future," Price pointed out.

DoD will relocate its affected systems to other bandwidths before December 2008, according to the Commerce Department.

Commerce officials said the 3G plan also calls for the private sector to gain another 45 MHz of bandwidth from the 2110-2170 MHz range, used by nongovernment entities.

The reallocation results from research and analysis made by commerce's NTIA, the FCC, DoD and other executive branch agencies.

Radio bandwidth is a fixed broadcast frequency spectrum divided up into ranges reserved for the public, governments, industry and various others. Commerce officials said the growing demand for advanced wireless services in the United States requires a larger share of bandwidth.

They noted that U.S. wireless use, measured in minutes, is increasing 75 percent annually. Therefore, adequate bandwidth is paramount for quality wireless voice and data service, to include cell phones and computerized communications systems used by U.S. industry and more than a hundred million consumers.

Arapahoe County
Colorado's First

J. Grayson Robinson
Sheriff

5686 South Court Place
Littleton, Colorado 80120
Phone: 303-734-5101
Fax: 303-734-5144
www.arapahoesheriff.org
sheriff@co.arapahoe.co.us

June 17, 2002

DOJ Grant Time Line & Equipment Purchase

U.S. Department of Justice, Office of State and Local Domestic Preparedness and Support, Office of Justice Program

In December 1998, it was decided that the law enforcement officers in Arapahoe County needed gas masks for personal protection. A grant was applied for and awarded in October of 1999 (1999-TE-CX-0001) for $45,360.00. This money was used to purchase Advantage 1000 CBA-RCA Gas Masks for 392 officers from 10 agencies in Arapahoe County and a Photo ionization Detector ($4200.00) for our Regional HAZMAT Team. The equipment was distributed in September of 2000.

Although this was a start, we still needed more. In 2001, Lt. Roger Hoffner, Arapahoe County Emergency Management Coordinator, Arapahoe County Sheriff's Office, applied for a grant to the Colorado Office of Emergency Management, who had the responsibility of distributing grant funds from the ***Department of Justice, Office of State and Local Domestic Preparedness and Support, "State Domestic Preparedness Equipment Program.***" The grant was a two-part process with a multi-county (Arapahoe and Douglas) Terrorist Assessment and the application, which listed the Arapahoe/Douglas, needs assessment at $731,720.00. In August of 2001, the Steering Committee preliminarily approved $352,000 for Arapahoe/Douglas Counties.

The process for the final approval and award from DOJ was to be sometime in November. Because of the subsequent terrorist attack on New York, the Arapahoe County Sheriff's Office, OEM, requested that $113,209.94 for personal protection equipment (PPE) ***"First Responder Bags"*** be expedited. The Colorado Office of Emergency Management passed along our concerns to DOJ and on September 26th the request was awarded (2EM70803). The money was used to purchase 805 ***"First Responder Bags"***. Because of the previously purchased gas masks in 2000, only 326 complete bags, with gas masks needed to be purchased. The rest were completed with the previously purchased masks.

The award notice for the balance of the $352,000 grant was received late May and is being approved and accepted through the County Commissioners. This will be used to enhance the regional Hazardous Material Team.

Respectfully,

Sheriff Patrick J. Sullivan, Jr.
Emergency Management Director

ACA
Accredited by the Commission on Accreditation for Law Enforcement Agencies, American Correctional Association and National Commission on Correctional Health Care.

The "First Responder Bags" contain:

♦ Advantage 1000 CW Gas Mask	$103.49
♦ Kappler CPF3 Level "B" Suit	$ 29.25
♦ Decontamination Kit (PIDS)	$ 26.95
♦ Duct Tape	$ 2.38
♦ Mono Goggles	$ 7.20
♦ Surgical Type gloves (4 mil Nitrile)	$ .22
♦ Green Nitrile Gloves	$ .95
♦ 1 ounce bottle of eye wash	$ 1.45
♦ (4) Disposal N95 Particulate respirators	$ 4.28
♦ (2) sets of ear plugs	$ .20
♦ (2) Latex over boots	$ 4.90
♦ 20 x 30 plastic bag for contaminated items	$ .50
♦ Canvas Storage Bag	$ 9.95
TOTAL COST	**$ 191.72**

<u>The bags were distributed to:</u>

Arapahoe County Sheriff's Office	208
Aurora Police Department	100
Bow Mar PD	2
Columbine Valley PD	4
Cherry Hills PD	23
Englewood PD	72
Greenwood PD	58
Littleton PD	64
Sheridan PD	21
Glendale PD	27
Douglas County Sheriff	125
Castle Rock PD	40
Parker PD	41
Public Works Cache	20
TOTAL	**805**

These bags will benefit Arapahoe Sheriff Office, Douglas County Sheriff's Office and 11 other law enforcement agencies in the counties. The response from the law enforcement community in getting these *"First responder Bags"* has been extremely positive.

Additionally, Arapahoe County Sheriff's Office purchased equipment to complete bags for: Coroners Office, Commissioners/Staff, Columbine Ambulance, Sheriff and Undersheriff, **Colorado State Patrol (32)** based in Castle Rock, and inventory to re-equip about 40 bags if used on an event.

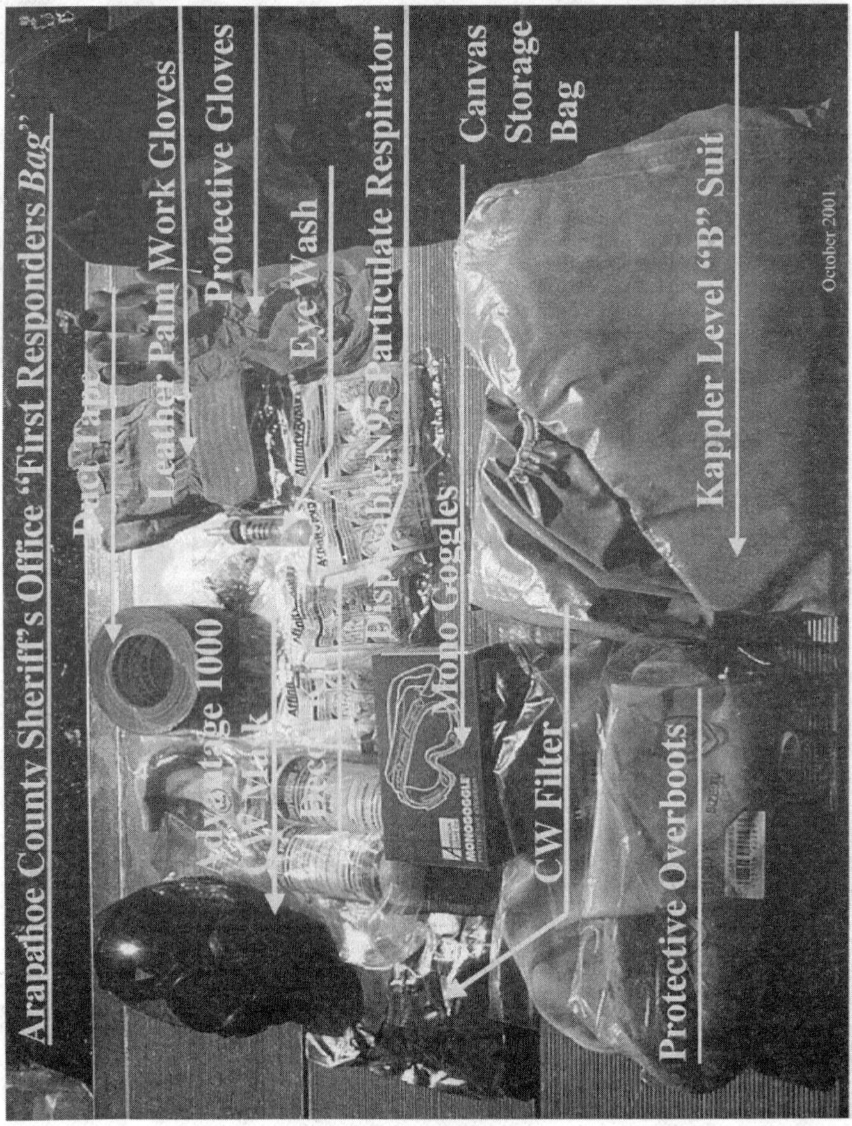
Arapahoe County Sheriff's Office "First Responders Bag"
Duct Tape
Leather Palm Work Gloves
Protective Gloves
Eye Wash
Disposable N95 Particulate Respirator
Canvas Storage Bag
Mono Goggles
CW Filter
Protective Overboots
Kappler Level "B" Suit
October 2001

Mr. HORN. Lieutenant Wicks, Office of Safety Services, Police Division, city of Englewood.

STATEMENT OF LIEUTENANT BYRON D. WICKS, OFFICE OF SAFETY SERVICES, POLICE DIVISION, CITY OF ENGLEWOOD

Mr. WICKS. Yes, sir. The Department of Safety Services for the city of Englewood incorporates fire and police as well as building codes and code enforcement. So when I speak, I speak both from the fire and police perspective.

My current assignment is as the investigations bureau commander. And before I get started, I want to make note that we are the benefactor of Lieutenant Hoffner's initiative and project. And we have, I believe, about 76 of those 805 bags that Lieutenant Hoffner's agency has donated to us, and we appreciate that.

Right after October or in October 2001, shortly after the September tragedy, my director appointed me czar of the Englewood Office of Preparedness and Security, as he called it, and Lieutenant Hoffner's been in this business for 6 years and if he feels confused; I've been in it for about 10 months and I know I am. So my perspective is definitely one from the operator's point of view, and that's based on 27 years of law enforcement experience, 31 years as a commissioned officer in the Marine Corps in which I was always assigned to operational billets. So as Lieutenant Hoffner indicates, operations at the level that we deal with is critical.

The two points that I want to bring up, and you're going to hear a lot of this, if you haven't already, is, first, the issue of interoperability. I know it's a hot topic, but if we don't have it, we are lost. And as sort of a subset to that issue of interoperability, I include communications, obviously, and that is definitely a problem, planning, and commonality in terms of response.

There isn't that much difference in a typical response that agencies in a suburban area would have. It's not to say that one size is going to fit all, but with minor modifications, it would be about a 95 percent solution. But when we are on one page and Arapahoe County's on another, even if we can talk to each other, we are not going to function well enough to do the job correctly the first time.

Which brings up the issue of exercises, both interagency and intra-agency exercises. And along with that, it would be nice if there was some standard, if you will, if Arapahoe County and Englewood had an exercise that maybe the State would say, "This is the way we want to go," so we have a common direction, a common focus, not just for our two agencies, but all agencies in the State of Colorado.

And then the second point I want to just address very quickly, and Roger virtually said everything I was going to say: As a grant administrator, funding issues are always an issue. One, obviously, is the amount of Federal funding in terms of grants, but maybe even more important than that, and Roger addressed this, is the fact that, one, how is it going to be spent? We recognize our own strengths and we also recognize our areas of need better than anybody else does.

Second of all is the form the grant comes in and the complexity that we have to deal with in terms of reporting what we've done with the moneys. Some grants, as a grant administrator, we will not accept, we will not apply for, because it's so difficult, it's so painful as an organization to account for that money, we simply won't do it we will not take any grants from the COPS program.

Mr. HORN. Did you ever have money from the COPS?

Mr. WICKS. Yes, we did, and we were audited, and it was incredibly painful, and it was just too difficult to prove what they were asking us to prove.

Mr. HORN. Well, you also had the problem of the localities and the cities of having to go out and add some people to their law enforcement, either their fire or for police and the sheriff's office, and that would be—got you out on a string there, and the Federal Government money suddenly comes off. And I don't know how much that was a factor in saying that—that program happened to be my law which was merged into it, because I wanted people just like you, a Marine, when you're retired from the military, I thought it was a good thing to do that because we needed police people. This was back in 1992–93, and so I was curious about what the problem was there with COPS. I do know there was a lot of bias as to the politics of it. That isn't unusual, but a little more than one.

Mr. WICKS. Well, we only spent about half our money and tried to give it back. That is impossible, to give back money on a grant. I mean, we couldn't spend it, we couldn't give it back. It was very burdensome. So right now the only grants that we will apply for are the block grants. And, quite frankly, I like the format of block grants because we get to choose the area in which we spend it, a one-page justification for the expenditures, and show them a receipt.

Mr. HORN. Well, I agree with you. I was a big advocate, beating the drums, long before I got into Congress in terms of revenue sharing, and that way the localities were much more able to figure out the needs than somebody sitting in Washington. That's my approach to it. There's the revenue sharing.

Mr. WICKS. And the last thing Roger brought up as well is we are one of the few police departments that uses the incident command system. Apparently Arapahoe County does. We use it because in safety services we combine with fire. Almost all fire departments use them but very few police departments. It's a great system, and it would be nice if the State agencies in the State of Colorado were like those in California where they all use the incident command system. They're all talking the same language and we all understand the responsibilities and jobs in a collective endeavor.

And with that, I'll rest.

Mr. HORN. That's very helpful, and we'll go in and do a few more things.

And now we have Paul L. Posner, Managing Director of Federal Budget Issues, Strategic Issues, U.S. General Accounting Office. Some of you, I think, are well known by it, and that is headed by the Comptroller General of the United States, currently Dave Walker, first-rate person, who believes in management and believes in people working together and all the rest. And he's got a 15-year appointment, and nobody can fuss around with him, including the President and the Congress, which is a pretty good deal. He deserves it, and he's doing a great job.

And so we've asked Mr. Posner to tell us what's missing and what haven't we got into that we should have gotten into. And if we've got something that isn't working right, we throw it over to him because we like his little blue books. There's about 50 blue books they've put out, by the way, on terrorism; isn't that right, easily 50?

Mr. POSNER. Probably more, but easily 50, yeah.

STATEMENT OF PAUL L. POSNER, MANAGING DIRECTOR, FEDERAL BUDGET ISSUES, STRATEGIC ISSUES, U.S. GENERAL ACCOUNTING OFFICE

Mr. POSNER. Well, thank you. It's a pleasure to be here, and it's a pleasure to hear the other testimonies such compelling ways showing that our Federal system is still vital. It reminds us at Washington how valiant the efforts have been out here, and when we craft new Federal responses, we better be sure that we support and don't kill that kind of initiative and passion.

We are realizing slowly that this challenge is beyond the capacity of any one level of government, including Washington, and it means that what we do here involves a national not a Federal response. It means it has to be collaborative, partnerial in nature. We have over 40 Federal agencies involved in this problem, 22 of which are going to be consolidated into the Department of Homeland Security, and there's still going to be a significant number of agencies not in this department, I might add. We have State governments from which we've heard, local governments, special districts; we have 87,000 units of those. We have private players who are critical in addressing this problem, and somehow we all have to figure out a way to integrate and overcome the stovepipes that have traditionally, at least at the Federal level, and are used to having these. I started my career with the New York City budget office and was familiar with stovepipes in city government as well.

One of the things that's so critical we've heard, particularly after September 11th, is the statement with regard to the first responders, and that's obviously a critical role. We've heard a lot about that, how you're better preparing yourself. But throughout the whole range of this problem, State and local governments are critical. But the last panel brought that up very well that even in counterintelligence and counterterrorism efforts, State and local governments are really critical players because we at the Federal level don't have the resources that you do. There are 650,000 police officers in this country, and it's dawning on Federal agencies that to respond to the kind of threat we are facing, they have to get into the community and have those kinds of information—they need to better find a way to tap the kind of information that you have and the resources you have. And that's why the INS is starting to contract with local police departments to chase down visa overstayers because their staffs simply aren't enough to do the job.

It's true for critical infrastructure. How do we protect critical infrastructure? The Federal Government doesn't own much and doesn't do much on its own. The critical infrastructure—the roads, the highways, the transit facilities, the ports in this country, the drinking water—are all really owned by State and local governments and private sector. And so how we can kind of figure out a way to mobilize a response on a national level is absolutely critical to solving this problem. And the dilemma for any local official, having been one myself at one time, is that you have really a lot of different players involved with these things. You don't control much, but, you sure are accountable for almost everything. And so that's why, you know, it's important that we help you better address those kinds of issues.

I will say that there are shifts going on right now in the way we—and we've heard here at the local level some of the important initiatives going on. At the Federal level, we all, of course, know about the President's proposed a Department of Homeland Security. The House has passed a bill largely following his proposal. The Senate has a somewhat different bill that has not yet passed. We are awaiting a resolution of that this fall. We believe at GAO that is a promising first step, a necessary but not sufficient step. It's important to get all the relevant players, for example, border security, get it together. Does that mean they'll all really work together? No. That's why putting them into a department is probably the first of a maybe 10-year journey before we really achieve the kind of harmonization and integration that we need. In the bill, for example, the Customs Service still has autonomy with regard to submitting its budget directly to Congress, notwithstanding the new department. So there are forces that still are going to be very difficult to address as the department, if we get a department, tries to bring some more cohesion to this.

Grants is one of the real important tools that we think the department will use and consolidation of grants is important. And you all have experienced the Department of Justice and FEMA and the Public Health Service and a variety of other separate funding streams coming down with different requirements, and we kind of dump it in your lap, and you've got to figure out how to bring them together. The Federal Government can do a better job of bringing some cohesion to that up front and providing some national goals, but giving you flexibility in how you address them.

So those are some of the challenges that we are starting to move to, but we've also seen some significant shifts already, not even passing the statute yet, in long-standing roles and relationships between these levels of government. National defense was historically a national responsibility, a Federal responsibility. Fire and police were historically a local responsibility. What this crisis is bringing to light is that defense is increasingly a local responsibility, protecting the Nation from this kind of insidious attack. And that, in fact, local police and how you work together is a national level interest. So we are bringing more of these kinds of levels that used to be separate together in some way, and the key is how can we do it in a way that both provides accountability to achieve some national expectations but gives flexibility to avoid a one-size-fits-all.

For example, when you look at what's in the offing right now, what's on the table, we have a new law that Congress passed requiring local drinking water systems to do vulnerability assessments and develop protection plans with some Federal money. We have a new Coast Guard regime that's putting a new Federal responsibility over the ports; that used to be a State and local responsibility. The Coast Guards requiring plans in 55 major ports. Fire services are again going to get a lot of Federal help but also probably some Federal standards. Communications, historically, we've heard, is a fairly fragmented thing. Every State and every community does it somewhat differently. The Office of Homeland Security's plans say that as a condition for Federal grants, local and State governments are going to try to achieve some greater interoperability, some national standards coming down in that arena. For the motor vehicles issue every State has a separate Department of Motor Vehicles. We are now seeing national proposals coming from Congress and the President to nationalize that responsibility, to get States to provide better secure documents when they issue driver's licenses so that we don't have terrorists able to counterfeit these kinds of documents. So there's going to be greater national pressures building on local services. The question is, how can we do it in a way that accommodates both national and local roles?

We've also seen State and local roles changing, and that's one of the issues when we talk about what's missing here. One of the things we've been doing is we've been going out, and we've seen, as jurisdictions within metropolitan regions attempting to work together more closely. Bringing together partners across a community as widespread as the Denver Metropolitan area is not an easy challenge. Every metropolitan area has tens, or hundreds in some cases, of governments, special purpose and others, that are responsible to their own constituencies. How to bring some harmonization together is always a challenge.

And so as we look down from the Federal angle, we see some of these problems are clearly local, some of them seem to be State, when the States can provide leadership, and some of them might best be done on a regional basis. And we've seen, as we visited, a lot of this starting to happen on its own. Mutual aid agreements historically have been in place. The public health networks are improving. And we are seeing a lot more here in Denver and other areas of regional efforts to promote better sharing and promote more economy of scale in how we provide for this expertise we need. But more, clearly, can be done in that regard, and it's historically very difficult to get communities, whether in the Washington area or New York area, to really collaborate with one another. And that's the kind of thing that we need to promote.

And we look at critical infrastructure areas, and we also see a lot of fragmentation there. Take, for example, airports. Airports have a patchwork quilt of different players responsible for security. We have TSA now responsible once you go through the gate. We have local and State governments responsible for the perimeter. We have the National Guard that comes in from time to time. We have the FAA that has responsibility. The airlines have responsibilities. So when we look at the safety of airlines and airline travel, we have a lot of players. And it's not clear to anyone that this has really been sorted out.

The same thing goes for something like food safety. You take the distribution chain and you have, you know, from the farm to the processor to the retail establishment to the grocery store or the restaurant, very different governmental roles and responsibilities for each stage of that process. The farmer is pretty much on their own. We have some kind of State roles there. When you get to the processing plants, why, there you have two different Federal agencies: The Agriculture Department responsible for meat and poultry, and the FDA responsible for pretty much everything else. We don't have any Federal standards; we have voluntary standards that those agencies have put out for that phase of the distribution process. When we get to the restaurants, that's totally State and local.

So what the President's homeland security plan suggested, and I think this is a useful thing to think about, is having a national strategy for each one of these critical infrastructure areas so that we would at least have a way to agree as a Nation whether we are comfortable and whether those roles and responsibilities are appropriate.

Given all these different players, it's really important to have clear goals and measures as we craft national strategies—what we are trying to achieve. How much security is enough, and how will we know it when we get there? The presence or absence of a terrorist event is not an acceptable performance measure. We want at national levels to make more of these investments. The quid pro quo is, I think, we are going to want to see some demonstrable changes in the results, in the outcomes. What are we getting by way of approved protection? Are there ways to measure it? Can we get every one of the systems to subscribe to those measures?

Finally, we've heard a lot of discussion, rightly so, about assistance, and we need to think more clearly at the national level about how we are going to get this done. Because, clearly, every hearing we do and every time we go to the local level, needs are incredibly large and always outrun the funding available. I'm reminded that the congressional budget office on Tuesday is going to issue their latest deficit update for the Federal Government; $160 billion deficit in 2002.

One of my other responsibilities at the General Accounting Office is to develop long-range Federal budget forecasts. And given the aging of our population and the increasing demands of Social Security and Medicare and Medicaid, why, our forecast of CBOs say that at current tax rates, we are going to be able to basically pay for the elderly and their doctors and that's about it in 20 years. In other words, we are quickly running, as the baby boomers retire, into dire fiscal straits. And so the question is, how do we respond to these urgent needs in a way that is both effective and economical? And that means we are going to have to think hard about how to best target these moneys, how to best ensure that we are going to get something of value for this. How, for example, to ensure that when we hand money down to local communities and States that they don't simply turn around and replace their own money with our money and cut taxes or put it in some other area. In other words, we need to prevent fiscal substitution. We need to have reasonable accountability provisions. I know that planning sometimes can go a little awry. Some kind of, again, assurance of results in terms of what we are getting for the money is important.

And, finally, there is the question of sustainability. How long should the Federal Government be involved, and what should be the Federal versus the State versus the local shares of costs in these things? So the point is, I think, by and large, we have to figure out a way to have a national and not a Federal approach. How do we balance accountability and flexibility, and how do we do it in a way that capitalizes on the strengths of each of the levels of government in forming a real partnership. Thank you.

[The prepared statement of Mr. Posner follows:]

United States General Accounting Office

GAO

Testimony

Before the Subcommittee on Government Efficiency, Financial Management, and Intergovernmental Relations, Committee on Government Reform, House of Representatives

For Release on Delivery
Expected at 10:00 a.m.,
Friday, August 23, 2002

HOMELAND SECURITY

Effective Intergovernmental Coordination Is Key to Success

Statement of Paul L. Posner, Managing Director,
Federal Budget Issues, Strategic Issues

GAO
Accountability * Integrity * Reliability

GAO-02-1013T

Mr. Chairman and Members of the Subcommittee:

I appreciate the opportunity to be here to discuss issues critical to successful federal leadership of, assistance to, and partnership with state and local governments to enhance homeland security. As you are aware, the challenges posed by homeland security exceed the capacity and authority of any one level of government. Protecting the nation against these unique threats calls for a truly integrated approach, bringing together the resources of all levels of government. The President's recently released national strategy for homeland security emphasizes security as a shared national responsibility involving close cooperation among all levels of government.[1] In addition, as you know, Mr. Chairman, the House has passed (H.R. 5005), and the Senate will take under consideration, after the August recess, legislation (S. 2452) to create a Department of Homeland Security. Although the bills are different, they share the goal of establishing a statutory Department of Homeland Security.

In my testimony today, I will focus on the challenges facing the federal government in (1) establishing a leadership structure for homeland security, (2) defining the roles of different levels of government, (3) developing performance goals and measures, and (4) deploying appropriate tools to best achieve and sustain national goals. My comments are based on a body of GAO's work on terrorism and emergency preparedness and policy options for the design of federal assistance,[2] our review of many other studies,[3] and the Comptroller General's recent testimonies on the proposed Department of Homeland Security (DHS).[4] In

[1] *National Strategy for Homeland Security.* The White House. Office of Homeland Security, July 16, 2002. In addition, the Office of Homeland Security issued a companion publication titled *State and Local Actions for Homeland Security* identifying measures state and local governments are taking to improve homeland security.

[2] See attached list of related GAO products.

[3] These studies include the Advisory Panel to Assess Domestic Response Capabilities for Terrorism Involving Weapons of Mass Destruction, *Third Annual Report* (Arlington, Va: Dec. 15, 2001); and the United States Commission on National Security/21st Century, *Road Map for Security: Imperative for Change* (February 15, 2001).

[4] *Homeland Security: Critical Design and Implementation Issues.* GAO-02-957T (Washington, D.C.: July 17, 2002) and *Homeland Security: Proposal for Cabinet Agency Has Merit, But Implementation Will Be Pivotal to Success.* GAO-02-886T (Washington, D.C.: June 25, 2002).

addition, I will draw on GAO's ongoing work for this Subcommittee, including an examination of the diverse ongoing and proposed federal preparedness programs, as well as a series of case studies we are conducting that examine preparedness issues facing state and local governments. To date, we have conducted interviews of officials in five geographically diverse cities: Baltimore, Maryland; Denver, Colorado; Los Angeles, California; New Orleans, Louisiana; and Seattle, Washington. We have also interviewed state emergency management officials in these states.

In summary:

- The proposed Department of Homeland Security will clearly have a central role in the success of efforts to enhance homeland security. Many aspects of a consolidation of homeland security programs have the potential to reduce fragmentation, improve coordination, and clarify roles and responsibilities. Realistically, however, in the short term, the magnitude of the challenges facing the new department will clearly require substantial time and effort and will take additional resources to make it effective. The recently released national strategy is intended to guide implementation of the complex mission of the proposed department and the efforts of other federal and non-federal entities responsible for homeland security initiatives.
- Appropriate roles and responsibilities within and between the levels of government and with the private sector are evolving and need to be clarified. New threats are prompting a reassessment and shifting of longstanding roles and responsibilities. Until now these shifts have been occurring on a piecemeal and ad hoc basis without benefit of an overarching framework and criteria to guide the process. The administration's national strategy recognizes the challenge posed by a complex structure of overlapping federal, state, and local governments—our country has more than 87,000 jurisdictions. There are also challenges in defining the appropriate roles and responsibilities of the private sector.
- The national strategy's initiatives often do not provide a baseline set of performance goals and measures upon which to assess and improve preparedness. Therefore, the nation does not yet have a comprehensive set of performance goals and measures upon which to assess and improve prevention efforts, vulnerability reduction, and responsiveness to damage and recovery needs at all levels of government. Given the need for a highly integrated approach to the homeland security challenge, national performance goals and measures for strategy initiatives that involve both

federal and non-federal actors may best be developed in a collaborative way involving all levels of government and the private sector. Standards are one tool the national strategy emphasizes in areas such as training, equipment, and communications.

- A careful choice of the most appropriate assistance tools is critical to achieve and sustain national goals. The choice and design of policy tools, such as grants, regulations, and tax incentives, can enhance the capacity of all levels of government to target areas of highest risk and greatest need, promote shared responsibilities by all parties, and track and assess progress toward achieving national preparedness goals. The national strategy notes that until recently, federal support for domestic preparedness efforts has been relatively small and disorganized, with various departments and agencies providing money in a "tangled web" of grant programs. It notes the shared responsibility of providing homeland security between federal, state, and local governments, and the private sector and recognizes the importance of using appropriate tools of government to improve preparedness.

Background

Homeland security is a complex mission that involves a broad range of functions performed throughout government, including law enforcement, transportation, food safety and public health, information technology, and emergency management, to mention only a few. Federal, state, and local governments have a shared responsibility in preparing for catastrophic terrorist attacks as well as other disasters. The initial responsibility for planning, preparing, and response falls upon local governments and their organizations—such as police, fire departments, emergency medical personnel, and public health agencies—which will almost invariably be the first responders to such an occurrence. For its part, the federal government has principally provided leadership, training, and funding assistance.

The federal government's role in responding to major disasters has historically been defined by the Stafford Act,[5] which makes most federal assistance contingent on a finding that the disaster is so severe as to be beyond the capacity of state and local governments to respond effectively. Once a disaster is declared, the federal government—through the Federal

[5] Robert T. Stafford Disaster Relief and Emergency Assistance Act (42 U.S.C. § 121 et seq.) establishes the process for states to request a presidential disaster declaration.

Emergency Management Agency (FEMA)—may reimburse state and local governments for between 75 and 100 percent of eligible costs, including response and recovery activities.

In addition to post disaster assistance, there has been an increasing emphasis over the past decade on federal support of state and local governments to enhance national preparedness for terrorist attacks. After the nerve gas attack in the Tokyo subway system on March 20, 1995, and the Oklahoma City bombing on April 19, 1995, the United States initiated a new effort to combat terrorism. In June 1995, Presidential Decision Directive 39 was issued, enumerating responsibilities for federal agencies in combating terrorism, including domestic terrorism. Recognizing the vulnerability of the United States to various forms of terrorism, the Congress passed the Defense Against Weapons of Mass Destruction Act of 1996 (also known as the Nunn-Lugar-Domenici program) to train and equip state and local emergency services personnel who would likely be the first responders to a domestic terrorist event. Other federal agencies, including those in FEMA; the departments of Justice, Health and Human Services, and Energy; and the Environmental Protection Agency, have also developed programs to assist state and local governments in preparing for terrorist events.

As emphasis on terrorism prevention and response grew, however, so did concerns over coordination and fragmentation of federal efforts. More than 40 federal entities have a role in combating and responding to terrorism, and more than 20 in bioterrorism alone. Our past work, conducted prior to the establishment of an Office of Homeland Security and the current proposals to create a new Department of Homeland Security, has shown coordination and fragmentation problems stemming largely from a lack of accountability within the federal government for terrorism-related programs and activities. Further, our work found there was an absence of a central focal point that caused a lack of a cohesive effort and the development of similar and potentially duplicative programs. Also, as the Gilmore Commission report notes, state and local officials have voiced frustration about their attempts to obtain federal funds from different programs administered by different agencies and have argued that the application process is burdensome and inconsistent among federal agencies.

President Bush has taken a number of important steps in the aftermath of the terrorist attacks of September 11th to address the concerns of fragmentation and to enhance the country's homeland security efforts, including creating of the Office of Homeland Security in October 2001, proposing the Department of Homeland Security in June 2002, and issuing a national strategy in July 2002. Both the House and Senate have worked diligently on these issues and are deliberating on a variety of homeland security proposals. The House has passed (H.R. 5005), and the Senate will take under consideration, after the August recess, legislation (S. 2452) to create a Department of Homeland Security. While these proposals would both transfer the functions, responsibilities, personnel, and other assets of existing agencies into the departmental structure, each bill has unique provisions not found in the other. For example, while both bills establish an office for State and Local Government Coordination and a first responder council to advise the department, the Senate bill also establishes a Chief Homeland Security Liaison Officer appointed by the Secretary and puts federal liaisons in each state to provide coordination between the department and the state and local first responders.

Proposed Department and National Strategy Will Guide Homeland Security

The proposal to create a statutorily based Department of Homeland Security holds promise to better establish the leadership necessary in the homeland security area. It can more effectively capture homeland security as a long-term commitment grounded in the institutional framework of the nation's governmental structure. As we have previously noted, the homeland security area must span the terms of various administrations and individuals. Establishing homeland security leadership by statute will ensure legitimacy, authority, sustainability, and the appropriate accountability to the Congress and the American people.[6]

The proposals call for the creation of a Cabinet department that would be responsible for coordination with other executive branch agencies involved in homeland security, including the Federal Bureau of Investigation and the Central Intelligence Agency. Additionally, the proposals call for coordination with nonfederal entities and direct the new Secretary to reach out to state and local governments and the private sector in order to: ensure adequate and integrated planning, training, and exercises occur, and that first responders have the necessary equipment; attaining interoperability of the federal government's homeland security communications systems with state and local governments' systems; oversee federal grant programs for state and local homeland security

[6]U.S. General Accounting Office, *Homeland Security: Responsibility And Accountability for Achieving National Goals.* GAO-02-627T (Washington, D.C.: Apr. 11, 2002).

efforts; and coordinate warnings and information to state and local government entities and the public.

Many aspects of the proposed consolidation of homeland security programs are in line with previous recommendations and show promise towards reducing fragmentation and improving coordination. For example, the new department would consolidate federal programs for state and local planning and preparedness from several agencies and place them under a single organizational umbrella. Based on our prior work, we believe that the consolidation of some homeland security functions makes sense and will, if properly organized and implemented, over time lead to more efficient, effective, and coordinated programs, better intelligence sharing, and a more robust protection of our people, borders, and critical infrastructure.

However, as the Comptroller General has recently testified,[7] implementation of the new department will be an extremely complex task, and in the short term, the magnitude of the challenges that the new department faces will clearly require substantial time and effort, and will take additional resources to make it effective. Further, some aspects of the new department, as proposed, may result in yet other concerns. For example, as we reported on June 25, 2002,[8] the new department could include public health assistance programs that have both basic public health and homeland security functions. These dual-purpose programs have important synergies that should be maintained and could potentially be disrupted by such a change.

The recently issued national strategy for homeland security states it is intended to answer four basic questions: what is "homeland security" and what missions does it entail; what does the nation seek to accomplish, and what are the most important goals of homeland security; what is the federal executive branch doing now to accomplish these goals and what should it do in the future; and what should non-federal governments, the

[7] U.S. General Accounting Office, *Homeland Security: Proposal for Cabinet Agency Has Merit, but Implementation Will Be Pivotal to Success*, GAO-02-886T (Washington, D.C.: June 25, 2002).

[8] U.S. General Accounting Office, *Homeland Security: New Department Could Improve Coordination but May Complicate Public Health Priority Setting*, GAO-02-883T (Washington, D.C.: June 25, 2002).

private sector, and citizens do to help secure the homeland. Within the federal executive branch, the key organization for homeland security will be the proposed Department of Homeland Security. The Department of Defense will contribute to homeland security, as well other departments such as the Departments of Justice, Agriculture, and Health and Human Services. The national strategy also makes reference to using tools of government such as grants and regulations to improve national preparedness.

The national strategy defines homeland security as a concerted national effort to 1) prevent terrorist attacks within the United States, 2) reduce America's vulnerability to terrorism, 3) minimize the damage, and 4) recover from attacks that do occur. This definition should help the government more effectively administer, fund, and coordinate activities both inside and outside the proposed new department and ensure all parties are focused on the same goals and objectives. The three parts of the definition form the national strategy's three objectives.

The strategy identifies six critical mission areas, and outlines initiatives in each of the six mission areas. It further describes four foundations that cut across these mission areas and all levels of government. These foundations— law; science and technology; information sharing and systems; and international cooperation— are intended to provide a basis for evaluating homeland security investments across the federal government. Table 1 summarizes key intergovernmental roles in each of the six mission areas as presented in the strategy.

Table 1: National Strategy: Six Critical Mission Areas and Key Intergovernmental Roles

Mission Area	Key Intergovernmental Roles
Intelligence and Warning	• Work with state and local law enforcement to leverage critical intelligence information, and provide real-time actionable information in the form of protective actions that should be taken in light of terrorist threats, trends, capabilities, and vulnerabilities. • Provide announcements of threat advisories and alerts to notify law enforcement and state and local government officials of threats through the Homeland Security Advisory System.
Border and Transportation Security	• Implementation of the Aviation and Transportation Security Act of 2001 requires partnerships among federal, state, and local government officials to assess and protect critical transportation infrastructures and reduce vulnerabilities.
Domestic Counterterrorism	• Expand data included in federal databases such as the FBI National Crime Information Center (NCIC) database and ensure that they are fully accessible to state and local law enforcement officials. • Expand the Joint Terrorism Task Forces, representing numerous federal agencies and state and local law enforcement, to all 56 FBI field offices.
Protecting Critical Infrastructures and Key Assets	• Work with state and local governments to implement a comprehensive national infrastructure protection plan to ensure protection for critical assets, systems, and functions, and for sharing protection responsibility with state and local government. • Provide state and local agencies one primary federal contact for coordinating protection activities with the federal government (e.g. vulnerability assessments, strategic planning efforts, and exercises).
Defending Against Catastrophic Threats	• In cooperation with state and local governments, develop additional inspection procedures and detection systems throughout the national transportation structure to detect the movement of nuclear materials within the U.S. • Expand and modernize the Centers for Disease Control Epidemic Intelligence Service to better train local and state officials in recognizing biological attacks, and state and local jurisdictions with a population of 500,000 or more will be provided with resources to hire skilled epidemiologists.
Emergency Preparedness and Response	• Working with state and local public safety organizations, build a comprehensive national incident management system to respond to terrorist incidents and natural disasters, and encourage first responder organizations to adopt the already widespread Incident Management System by making it a requirement for federal grants. • Provide grants in support of state and local preparedness efforts in areas such as: mutual aid agreements; terrorism-related communications equipment; training and equipping of state and local health care personnel to deal with chemical, biological, radiological, and nuclear terrorism; planning for the receipt and distribution of medicines from the National Pharmaceutical Stockpile; equipping, training, and exercising first responders to meet certification standards. • Proposed grant requirements include: compliance with a national emergency communication plan, progress in achieving communications interoperability with other emergency response bodies, and annual certification of first responder preparedness to handle and decontaminate any hazard. • Consolidate all grant programs that distribute federal funds to state and local first responders. The First Responder Initiative proposes to increase federal funding levels more than tenfold to $3.5 billion in fiscal year 2003.

With regard to the costs of Homeland Security, the national strategy emphasizes government should fund only those homeland security activities that are not supplied, or are inadequately supplied, in the market, and cost sharing between different governmental levels should reflect federalism principles and different tools of government. In terms of the financial contributions made by state and local government to homeland security, the strategy acknowledges that state and local governments are incurring unexpected costs defending or protecting their respective communities. These costs include protecting critical infrastructure, improving technologies for information sharing and communications, and building emergency response capacity. At this time, the National Governors' Association estimates that additional homeland security-related costs, incurred since September 11th and through the end of 2002, will reach approximately $6 billion. Similarly, the U.S. Conference of Mayors has estimated the costs incurred by cities during this time period to be $2.6 billion.

Challenges Remain in Defining Appropriate Intergovernmental Roles

The proposed department will be a key player in the daunting challenge of defining the roles of the various actors within the intergovernmental system responsible for homeland security. In areas ranging from fire protection to drinking water to port security, the new threats are prompting a reassessment and shift of longstanding roles and responsibilities. However, until this time, proposed shifts in roles and responsibilities have been considered on a piecemeal and ad hoc basis without benefit of an overarching framework and criteria to guide this process. The national strategy recognizes that the process is challenging because of the structure of overlapping federal, state, and local governments given that our country has more than 87,000 jurisdictions. The national strategy further notes that the challenge is to develop interconnected and complementary systems that are reinforcing rather than duplicative.

The proposals for a Department of Homeland Security call for the department to reach out to state and local governments and the private sector to coordinate and integrate planning, communications, information, and recovery efforts addressing homeland security. This is important recognition of the critical role played by nonfederal entities in protecting the nation from terrorist attacks. State and local governments play primary roles in performing functions that will be essential to effectively address our new challenges. Much attention has already been paid to their role as first responders in all disasters, whether caused by terrorist attacks or natural hazards.

The national strategy emphasizes the critical role state and local governments play in homeland security and the need for coordination between all levels of government. The national strategy emphasizes that homeland security is a shared responsibility. In addition, the national strategy has several initiatives designed to improve partnerships and coordination. Table 1 provides several examples of areas with key intergovernmental roles and coordination. For example, there are initiatives to improve intergovernmental law enforcement coordination and enabling effective partnerships with state and local governments and the private sector in critical infrastructure protection. States are asked to take several legal initiatives, such as coordinating suggested minimum standards for state driver's licenses and reviewing quarantine authorities. Many initiatives are intended to develop or enhance first responder capabilities, such as initiatives to improve the technical capabilities of first responders or enable seamless communication among all responders. In many cases, these initiatives will rely on federal, state, and local cooperation, some standardization, and the sharing of costs.

National and Regional Partnerships

Achieving national preparedness and response goals hinges on the federal government's ability to form effective partnerships with nonfederal entities. Therefore, federal initiatives should be conceived as national, not federal in nature. Decision makers have to balance the national interest of prevention and preparedness with the unique needs and interests of local communities. A "one-size-fits-all" federal approach will not serve to leverage the assets and capabilities that reside within state and local governments and the private sector. By working collectively with state and local governments, the federal government gains the resources and expertise of the people closest to the challenge. For example, protecting infrastructure such as water and transit systems lays first and most often with nonfederal levels of government.

Just as partnerships offer opportunities, they also pose risks based upon the different interests reflected by each partner. From the federal perspective, there is the concern that state and local governments may not share the same priorities for use of federal funds. This divergence of priorities can result in state and local governments simply replacing ("supplanting") their own previous levels of commitment in these areas with the new federal resources. From the state and local perspective, engagement in federal programs opens them up to potential federal preemption and mandates. From the public's perspective, partnerships if not clearly defined, risk blurring responsibility for the outcome of public programs.

Our fieldwork at federal agencies and at local governments suggests a shift is potentially underway in the definition of roles and responsibilities between federal, state, and local governments with far reaching consequences for homeland security and accountability to the public. The challenges posed by the new threats are prompting officials at all levels of government to rethink long-standing divisions of responsibilities for such areas as fire services, local infrastructure protection, and airport security. Current homeland security proposals recognize that the unique scale and complexity of these threats call for a response that taps the resources and capacities of all levels of government as well as the private sector.

In many areas, these proposals would impose a stronger federal presence in the form of new national standards or assistance. For instance, the Congress is considering proposals to mandate new vulnerability assessments and protective measures on local communities for drinking water facilities. Similarly, new federal rules have mandated local airport authorities to provide new levels of protection for security around airport perimeters. The block grant proposal for first responders would mark a dramatic upturn in the magnitude and role of the federal government in providing assistance and standards for fire service training and equipment.

Additionally, the national strategy suggests initiatives for an expanded state role in several areas. For example, there are no national or agreed upon state standards for driver's license content, format, or acquisition procedures. The strategy states that the federal government should support state-led efforts to develop suggested minimum standards for drivers' licenses. In another example, in order to suppress money laundering, the strategy recommends that states assess the current status of their regulation regarding providers of financial services and work to adopt uniform laws as necessary.

Governments at the local level are also moving to rethink roles and responsibilities to address the unique scale and scope of the contemporary threats from terrorism. Numerous local general-purpose governments and special districts co-exist within metropolitan regions and rural areas alike. Many regions are starting to assess how to restructure relationships among contiguous local entities to take advantage of economies of scale, promote resource sharing, and improve coordination of preparedness and response on a regional basis. In our case studies of five metropolitan areas, we have identified several common forms of regional cooperation and coordination including special task forces or working groups, improved collaboration among public health entities, increased countywide planning, mutual aid agreements, and communications. These partnerships are at varying stages of development and are continuing to evolve. Table 2 summarizes these initiatives.

Table 2: Case Study Examples of Metropolitan Cooperation and Coordination

- **Task Forces and Working Groups:** To facilitate emergency planning and coordination among cities in a metropolitan area, officials have joined together to create task forces, such as terrorism working groups, advisory committees, and Mayors' caucuses. For example, the Metropolitan Safety, Security, and Anti-terrorism Task Force in New Orleans includes officials from the city and four surrounding parishes.
- **Collaboration with Public Health Entities:** Public health departments, emergency medical services, and hospitals are participating in planning efforts to coordinate use of limited resources such as emergency room capacity, hospital beds, and medical supplies. For example, in Denver, the Front Range Emergency Medical Service and Trauma Advisory Council involves all hospitals and rescue squads in a six-county metropolitan area.
- **Countywide Planning:** In some states, counties serve as the primary coordinating agent and work with cities within their jurisdiction, other counties, and the state to ensure that they develop and update emergency and disaster plans, provide training, conduct assessments and exercises, and have adequate emergency resources. For example, King County, Washington has coordinated development of a Regional Disaster Plan, which includes Seattle and 15 other cities within the county as well as 15 fire districts, 15 hospitals, 21 water and sewer districts, 12 school districts, and the private sector.
- **Mutual Aid Agreements:** Cities and counties have used mutual aid agreements to share emergency resources in their metropolitan areas. These agreements may include fire, police, emergency medical services, and hospitals and may be formal or informal. For example, Los Angeles has mutual aid agreements between police and fire departments in surrounding jurisdictions and a range of private sector entities. The state has a Mutual Aid Regional Advisory Commission that facilitates agreements, and the Standardized Emergency Management System (SEMS) law requires mutual aid agreements for state reimbursement.
- **Communications:** Cities and counties currently use a variety of methods for communicating among first responders, such as command centers, using radio, cell phones, and pagers; amateur radio operators; and community alert systems. Some are considering 800 MHz radio systems to permit interoperability and mobile incident command centers to direct communications among first responders. King County, Washington has a countywide 800 MHz system and uses amateur radio operators to provide a redundant emergency communications system.

Although promising greater levels of protection than before, these shifts in roles and responsibilities have been developed on an ad hoc piecemeal basis without the benefit of common criteria. An ad hoc process may not capture the real potential each actor in our system offers. Moreover, a piecemeal redefinition of roles risks the further fragmentation of the responsibility for homeland security within local communities, blurring lines of responsibility and accountability for results. While federal, state, and local governments all have roles to play, care must be taken to clarify who is responsible for what so that the public knows whom to contact to address their problems and concerns. Current homeland security initiatives provide an opportunity to more systematically identify the unique resources and capacities of each level of government and better match these capabilities to the particular tasks at hand. If implemented in a partnerial fashion, the national strategy can also promote the participation, input, and buy in of state and local partners whose cooperation is essential for success.

Performance Goals and Measures Needed in Homeland Security Programs

The proposed department, in fulfilling its broad mandate, has the challenge of developing a national performance focus. The national strategy is a good start in defining strategic objectives and related mission areas, plus foundations that cut across the mission areas. The national strategy's initiatives to implement the objectives under the related mission and foundation areas extend from building capabilities to achieving specific outcomes.

According to the national strategy, each department and agency is to be held accountable for its performance on homeland security efforts. However, the initiatives often do not provide a baseline set of goals and measures upon which to assess and improve many of its initiatives to prevent attacks, reduce the nation's vulnerability to attacks, or minimize the damage and recovering from attacks that do occur. For example, the initiative of creating "smart borders" requires a clear specification of what is expected of a smart border, including consideration of security and economic aspects of moving people and goods.

Specific performance goals and measures for many initiatives will occur at a later date. The strategy states that each department or agency will create benchmarks and other performance measures to evaluate progress and allocate future resources. Performance measures will be used to evaluate the effectiveness of each homeland security program, allowing agencies to measure their progress, make resource allocation decisions, and adjust priorities. As the national strategy and related implementation plans evolve, we would expect clearer performance expectations to emerge. Given the need for a highly integrated approach to the homeland security challenge, national performance goals and measures may best be developed in a collaborative way involving all levels of government and the private sector.

Assessing the capability of state and local governments to respond to catastrophic terrorist attacks is an important feature of the national strategy and the responsibilities of the proposed new department. The President's fiscal year 2003 budget proposal acknowledged that our capabilities for responding to a terrorist attack vary widely across the country. The national strategy recognizes the importance of standards and performance measures in areas such as training, equipment, and communications. For example, the national strategy proposes the establishment of national standards for emergency response training and preparedness. These standards would require certain coursework for individuals to receive and maintain certification as first responders and for state and local governments to receive federal grants. Under the strategy, the proposed department would establish a national exercise program designed to educate and evaluate civilian response personnel at all levels of government. It would require individuals and government bodies to complete successfully at least one exercise every year. The department would use these exercises to measure performance and allocate future resources.

Standards are being developed in other areas associated with homeland security, yet formidable challenges remain. For example, national standards that would apply to all ports and all public and private facilities are well under way. In preparing to assess security conditions at 55 U.S. ports, the Coast Guard's contractor has been developing a set of standards since May 2002. These standards cover such things as preventing unauthorized persons from accessing sensitive areas, detecting and intercepting intrusions, and checking backgrounds of those whose jobs require access to port facilities. However, challenges remain in finalizing a complete set of standards for the level of security needed in the nation's ports, resolving issues between key stakeholders that have conflicting or competing interests, and establishing mechanisms for enforcement. Moreover, because security at ports is a concern shared among federal, state, and local governments, as well as among private commercial interests, the issue of who should pay to finance antiterrorism activities may be difficult to resolve.

Communications is an example of an area for which standards have not yet been developed, but various emergency managers and other first responders have continuously highlighted that standards are needed. State and local governments often report that there are deficiencies in their communications capabilities, including the lack of interoperable systems. The national strategy recognizes that it is crucial for response personnel to have and use equipment, systems, and procedures that allow them to communicate. Therefore, the strategy calls for the proposed Department of Homeland Security to develop a national communication plan to establish protocols (who needs to talk to whom), processes, and national standards for technology acquisition. According to the national strategy, this is a priority for fiscal year 2003 funding which ties all federal grant programs that support state and local purchase of terrorism-related communications equipment to this communication plan.

The establishment of specific national goals and measures for homeland security initiatives, including preparedness, will not only go a long way towards assisting state and local entities in determining successes and areas where improvement is needed, but could also be used as goals and performance measures as a basis for assessing the effectiveness of federal programs. The Administration should take advantage of the Government Performance and Results Act (GPRA) and its performance tools of strategic plans, annual performance plans and measures, and accountability reports for homeland security implementation planning. At the department and agency level, until the new department is operational, GPRA can be a useful tool in developing homeland security implementation plans within and across federal agencies. Given the recent and proposed increases in homeland security funding, as well as the need for real and meaningful improvements in preparedness, establishing clear goals and performance measures is critical to ensuring both a successful and fiscally responsible effort.

Appropriate Tools Need to Be Selected for Providing Assistance

The choice and design of the policy tools the federal government uses to engage and involve other levels of government and the private sector in enhancing homeland security will have important consequences for performance and accountability. Governments have a variety of policy tools including grants, regulations, tax incentives, and information-sharing mechanisms to motivate or mandate other levels of government or the private sector to address security concerns. The choice of policy tools will affect sustainability of efforts, accountability and flexibility, and targeting of resources. The design of federal policy will play a vital role in determining success and ensuring that scarce federal dollars are used to achieve critical national goals. The national strategy acknowledges the shared responsibility of providing homeland security between federal, state, and local governments, and the private sector and recognizes the importance of using tools of government such as grants, regulations, and information sharing to improve national preparedness.

Grants

The federal government often uses grants to state and local governments as a means of delivering federal assistance. Categorical grants typically permit funds to be used only for specific, narrowly defined purposes. Block grants typically can be used by state and local governments to support a range of activities aimed at achieving a broad, national purpose and to provide a great deal of discretion to state and local officials. In designing grants, it is important to (1) target the funds to states and localities with the greatest need based on highest risk and lowest capacity to meet these needs from their own resource bases, (2) discourage the replacement of state and local funds with federal funds, commonly referred to as supplantation, with a maintenance-of-effort requirement that recipients maintain their level of previous funding, and (3) strike a balance between accountability and flexibility. At their best, grants can stimulate state and local governments to enhance their preparedness to address the unique threats posed by terrorism. Ideally, grants should stimulate higher levels of preparedness and avoid simply subsidizing local functions that are traditionally state or local responsibilities. One approach used in other areas is the "seed money" model in which federal grants stimulate initial state and local activity with the intent of transferring responsibility for sustaining support over time to state and local governments.

Recent funding proposals, such as the $3.5 billion block grant for first responders contained in the president's fiscal year 2003 budget, have included some of these provisions. This grant would be used by state and local governments to purchase equipment; train personnel; and exercise, develop, or enhance response plans. Once the details of the grant have been finalized, it will be useful to examine the design to assess how well the grant will target funds, discourage supplantation, and provide the appropriate balance between accountability and flexibility, and whether it provides temporary "seed money" or represents a long-term funding commitment.

Regulations

Other federal policy tools can also be designed and targeted to elicit a prompt, adequate, and sustainable response. In the area of regulatory authority, the federal, state, and local governments share authority for setting standards through regulations in several areas, including infrastructure and programs vital to preparedness (for example, transportation systems, water systems, and public health). In designing regulations, key considerations include how to provide federal protections, guarantees, or benefits while preserving an appropriate balance between federal and state and local authorities and between the public and private sectors. Regulations have recently been enacted in the area of infrastructure. For example, a new federal mandate requires that local drinking water systems in cities above a certain size provide a vulnerability assessment and a plan to remedy vulnerabilities as part of ongoing EPA reviews, while the Transportation and Aviation Security Act grants the Department of Transportation authority to order deployment of local law enforcement personnel in order to provide perimeter access security at the nation's airports.

In designing a regulatory approach, the challenges include determining who will set the standards and who will implement or enforce them.

Several models of shared regulatory authority offer a range of approaches that could be used in designing standards for preparedness. Examples of these models range from preemption through fixed federal standards to state and local adoption of voluntary standards formulated by quasi-official or nongovernmental entities.[9]

Tax Incentives

As the administration noted, protecting America's infrastructure is a shared responsibility of federal, state, and local government, in active partnership with the private sector, which owns approximately 85 percent of our nation's critical infrastructure. To the extent that private entities will be called upon to improve security over dangerous materials or to protect critical infrastructure, the federal government can use tax incentives to encourage or enforce their activities. Tax incentives are the result of special exclusions, exemptions, deductions, credits, deferrals, or tax rates in the federal tax laws. Unlike grants, tax incentives do not generally permit the same degree of federal oversight and targeting, and they are generally available by formula to all potential beneficiaries who satisfy congressionally established criteria.

Information Sharing

Since the events of September 11th, a task force of mayors and police chiefs has called for a new protocol governing how local law enforcement agencies can assist federal agencies, particularly the FBI. As the U.S. Conference of Mayors noted, a close working partnership of federal and local law enforcement agencies, which includes the sharing of information, will expand and strengthen the nation's overall ability to prevent and respond to domestic terrorism. The USA Patriot Act provides for greater sharing of information among federal agencies. An expansion of this act has been proposed (S1615; H.R. 3285) that would provide for information sharing among federal, state, and local law enforcement agencies. In addition, the Intergovernmental Law Enforcement Information Sharing Act of 2001 (H.R. 3483), which you sponsored, Mr. Chairman, addresses a number of information-sharing needs. For instance, the proposed legislation provides that the Attorney General expeditiously grant security clearances to Governors who apply for them and to state and local officials who participate in federal counterterrorism working groups or regional task forces.

[9]For more information on these models, see U.S. General Accounting Office, *Regulatory Programs: Balancing Federal and State Responsibilities for Standard Setting and Implementation.* GAO-02-495 (Washington, D.C.: March 20, 2002).

The national strategy also includes several information-sharing and systems initiatives to facilitate dissemination of information from the federal government to state and local officials. For example, the strategy supports building and sharing law enforcement databases, secure computer networks, secure video teleconferencing capabilities, and more accessible websites. It also states that the federal government will make an effort to remove classified information from some documents to facilitate distribution to more state and local authorities.

Conclusion

The recent publication of the national strategy is an important initial step in defining homeland security, setting forth key strategic objectives, and specifying initiatives to implement them. The proposals for the Department of Homeland Security represent recognition by the administration and the Congress that much still needs to be done to improve and enhance the security of the American people and our country's assets. The proposed department will clearly have a central role in the success of efforts to strengthen homeland security, and has primary responsibility for many of the initiatives in the national homeland security strategy.

Moreover, given the unpredictable characteristics of terrorist threats, it is essential that the strategy be implemented at a national rather than federal level with specific attention given to the important and distinct roles of state and local governments. Accordingly, decision makers will have to balance the federal approach to promoting homeland security with the unique needs, capabilities, and interests of state and local governments. Such an approach offers the best promise for sustaining the level of commitment needed to address the serious threats posed by terrorism.

This completes my prepared statement. I would be pleased to respond to any questions you or other Members of the Subcommittee may have.

Contacts and Acknowledgments

For further information about this testimony, please contact me at (202) 512-9573 or JayEtta Hecker at (202) 512-2834. Other key contributors to this testimony include Matthew Ebert, Thomas James, David Laverny-Rafter, Yvonne Pufahl, Jack Schulze, and Amelia Shachoy.

Related GAO Products

Homeland Security

Port Security: Nation Faces Formidable Challenges in Making New Initiatives Successful. GAO-02-993T. Washington, D.C.: August 5, 2002.

Aviation Security: Transportation Security Administration Faces Immediate and Long-Term Challenges. GAO-02-971T. Washington, D.C.: July 25, 2002.

Homeland Security: Critical Design and Implementation Issues. GAO-02-957T. Washington, D.C.: July 17, 2002.

Homeland Security: New Department Could Improve Coordination but Transferring Control of Certain Public Health Programs Raises Concerns. GAO-02-954T. Washington, D.C.: July 16, 2002.

Critical Infrastructure Protection: Significant Homeland Security Challenges Need to Be Addressed. GAO-02-918T. Washington, D.C.: July 9, 2002.

Homeland Security: New Department Could Improve Biomedical R&D Coordination but May Disrupt Dual-Purpose Efforts. GAO-02-924T. Washington, D.C.: July 9, 2002.

Homeland Security: Title III of the Homeland Security Act of 2002. GAO-02-927T. Washington, D.C.: July 9, 2002.

Homeland Security: Intergovernmental Coordination and Partnership Will Be Critical to Success. GAO-02-901T. Washington, D.C.: July 3, 2002.

Homeland Security: New Department Could Improve Coordination but May Complicate Priority Setting. GAO-02-893T. Washington, D.C.: June 28, 2002.

Homeland Security: New Department Could Improve Coordination but May Complicate Public Health Priority Setting. GAO-02-883T. Washington, D.C.: June 25, 2002.

Homeland Security: Proposal for Cabinet Agency Has Merit, But Implementation Will Be Pivotal to Success. GAO-02-886T. Washington, D.C.: June 25, 2002.

Homeland Security: Key Elements to Unify Efforts Are Underway but Uncertainty Remains. GAO-02-610. Washington, D.C.: June 7, 2002.

National Preparedness: Integrating New and Existing Technology and Information Sharing into an Effective Homeland Security Strategy. GAO-02-811T. Washington, D.C.: June 7, 2002.

Homeland Security: Integration of Federal, State, Local, and Private Sector Efforts Is Critical to an Effective National Strategy for Homeland Security GAO-02-621T. Washington, D.C.: April 11, 2002.

Combating Terrorism: Enhancing Partnerships Through a National Preparedness Strategy. GAO-02-549T. Washington, D.C.: March 28, 2002.

Homeland Security: Progress Made, More Direction and Partnership Sought. GAO-02-490T. Washington, D.C.: March 12, 2002.

Homeland Security: Challenges and Strategies in Addressing Short- and Long-Term National Needs. GAO-02-160T. Washington, D.C.: November 7, 2001.

Homeland Security: A Risk Management Approach Can Guide Preparedness Efforts. GAO-02-208T. *Washington, D.C.: October 31, 2001.*

Homeland Security: Need to Consider VA's Role in Strengthening Federal Preparedness. GAO-02-145T. Washington, D.C.: October 15, 2001.

Homeland Security: Key Elements of a Risk Management Approach. GAO-02-150T. Washington, D.C.: October 12, 2001.

Homeland Security: A Framework for Addressing the Nation's Issues. GAO-01-1158T. Washington, D.C.: September 21, 2001.

Combating Terrorism

Combating Terrorism: Intergovernmental Cooperation in the Development of a National Strategy to Enhance State and Local Preparedness. GAO-02-550T. Washington, D.C.: April 2, 2002.

Combating Terrorism: Enhancing Partnerships Through a National Preparedness Strategy. GAO-02-549T. Washington, D.C.: March 28, 2002.

Combating Terrorism: Critical Components of a National Strategy to Enhance State and Local Preparedness. GAO-02-548T. Washington, D.C.: March 25, 2002.

Combating Terrorism: Intergovernmental Partnership in a National Strategy to Enhance State and Local Preparedness. GAO-02-547T. Washington, D.C.: March 22, 2002.

Combating Terrorism: Key Aspects of a National Strategy to Enhance State and Local Preparedness. GAO-02-473T. Washington, D.C.: March 1, 2002.

Combating Terrorism: Considerations for Investing Resources in Chemical and Biological Preparedness. GAO-01-162T. Washington, D.C.: October 17, 2001.

Combating Terrorism: Selected Challenges and Related Recommendations. GAO-01-822. Washington, D.C.: September 20, 2001.

Combating Terrorism: Actions Needed to Improve DOD's Antiterrorism Program Implementation and Management. GAO-01-909. Washington, D.C.: September 19, 2001.

Combating Terrorism: Comments on H.R. 525 to Create a President's Council on Domestic Preparedness. GAO-01-555T. Washington, D.C.: May 9, 2001.

Combating Terrorism: Observations on Options to Improve the Federal Response. GAO-01-660T. Washington, D.C.: April 24, 2001.

Combating Terrorism: Comments on Counterterrorism Leadership and National Strategy. GAO-01-556T. Washington, D.C.: March 27, 2001.

Combating Terrorism: FEMA Continues to Make Progress in Coordinating Preparedness and Response. GAO-01-15. Washington, D.C.: March 20, 2001.

Combating Terrorism: Federal Response Teams Provide Varied Capabilities; Opportunities Remain to Improve Coordination. GAO-01-14. Washington, D.C.: November 30, 2000.

Combating Terrorism: Need to Eliminate Duplicate Federal Weapons of Mass Destruction Training. GAO/NSIAD-00-64. Washington, D.C.: March 21, 2000.

Combating Terrorism: Observations on the Threat of Chemical and Biological Terrorism. GAO/T-NSIAD-00-50. Washington, D.C.: October 20, 1999.

Combating Terrorism: Need for Comprehensive Threat and Risk Assessments of Chemical and Biological Attack. GAO/NSIAD-99-163. Washington, D.C.: September 7, 1999.

Combating Terrorism: Observations on Growth in Federal Programs. GAO/T-NSIAD-99-181. Washington, D.C.: June 9, 1999.

Combating Terrorism: Analysis of Potential Emergency Response Equipment and Sustainment Costs. GAO-NSIAD-99-151. Washington, D.C.: June 9, 1999.

Combating Terrorism: Use of National Guard Response Teams Is Unclear. GAO/NSIAD-99-110. Washington, D.C.: May 21, 1999.

Combating Terrorism: Observations on Federal Spending to Combat Terrorism. GAO/T-NSIAD/GGD-99-107. Washington, D.C.: March 11, 1999.

Combating Terrorism: Opportunities to Improve Domestic Preparedness Program Focus and Efficiency. GAO-NSIAD-99-3. Washington, D.C.: November 12, 1998.

Combating Terrorism: Observations on the Nunn-Lugar-Domenici Domestic Preparedness Program. GAO/T-NSIAD-99-16. Washington, D.C.: October 2, 1998.

Combating Terrorism: Threat and Risk Assessments Can Help Prioritize and Target Program Investments. GAO/NSIAD-98-74. Washington, D.C.: April 9, 1998.

Combating Terrorism: Spending on Governmentwide Programs Requires Better Management and Coordination. GAO/NSIAD-98-39. Washington, D.C.: December 1, 1997.

Public Health

Homeland Security: New Department Could Improve Coordination but may Complicate Public Health Priority Setting. GAO-02-883T. Washington, D.C.: June 25, 2002.

Bioterrorism: The Centers for Disease Control and Prevention's Role in Public Health Protection. GAO-02-235T. Washington, D.C.: November 15, 2001.

Bioterrorism: Review of Public Health and Medical Preparedness. GAO-02-149T. Washington, D.C.: October 10, 2001.

Bioterrorism: Public Health and Medical Preparedness. GAO-02-141T. Washington, D.C.: October 10, 2001.

Bioterrorism: Coordination and Preparedness. GAO-02-129T. Washington, D.C.: October 5, 2001.

Bioterrorism: Federal Research and Preparedness Activities. GAO-01-915. Washington, D.C.: September 28, 2001.

Chemical and Biological Defense: Improved Risk Assessments and Inventory Management Are Needed. GAO-01-667. Washington, D.C.: September 28, 2001.

West Nile Virus Outbreak: Lessons for Public Health Preparedness. GAO/HEHS-00-180. Washington, D.C.: September 11, 2000.

Need for Comprehensive Threat and Risk Assessments of Chemical and Biological Attacks. GAO/NSIAD-99-163. Washington, D.C.: September 7, 1999.

Chemical and Biological Defense: Program Planning and Evaluation Should Follow Results Act Framework. GAO/NSIAD-99-159. Washington, D.C.: August 16, 1999.

Combating Terrorism: Observations on Biological Terrorism and Public Health Initiatives. GAO/T-NSIAD-99-112. Washington, D.C.: March 16, 1999.

Disaster Assistance

Disaster Assistance: Improvement Needed in Disaster Declaration Criteria and Eligibility Assurance Procedures. GAO-01-837. Washington, D.C.: August 31, 2001.

FEMA and Army Must Be Proactive in Preparing States for Emergencies. GAO-01-850. Washington, D.C.: August 13, 2001.

Federal Emergency Management Agency: Status of Achieving Key Outcomes and Addressing Major Management Challenges. GAO-01-832. Washington, D.C.: July 9, 2001.

Budget and Management

Managing for Results: Progress in Linking Performance Plans with Budget and Financial Statements. GAO-02-236. Washington, D.C.: January 4, 2002.

Results-Oriented Budget Practices in Federal Agencies. GAO-01-1084SP. Washington, D.C.: August 2001.

Managing for Results: Federal Managers' Views on Key Management Issues Vary Widely across Agencies. GAO-01-0592. Washington, D.C.: May 2001.

Determining Performance and Accountability Challenges and High Risks. GAO-01-159SP. Washington, D.C.: November 2000.

Managing for Results: Using the Results Act to Address Mission Fragmentation and Program Overlap. GAO/AIMD-97-156. Washington, D.C.: August 29, 1997.

Government Restructuring: Identifying Potential Duplication in Federal Missions and Approaches. GAO/T-AIMD-95-161. Washington, D.C.: June 7, 1995.

Grant Design

Grant Programs: Design Features Shape Flexibility, Accountability, and Performance Information. GAO/GGD-98-137. Washington, D.C.: June 22, 1998.

Federal Grants: Design Improvements Could Help Federal Resources Go Further. GAO/AIMD-97-7. Washington, D.C.: December 18, 1996.

Block Grants: Issues in Designing Accountability Provisions. GAO/AIMD-95-226. Washington, D.C.: September 1, 1995.

Mr. HORN. OK. And let us go back a minute now on a few questions, and then we'll wrap it up.

Mr. Wall, I was curious as to what degree are the Veterans Administration hospitals and clinics working with your overall association, and what do you feel the VA can do in terms of some of the things we've talked about in terms of the attacks of chemistry, biology, you name it, and are we prepared for the private, nonprofits, and the VA, and the military hospitals?

Mr. WALL. Several questions there. I'm not sure I have answers to all of those, but——

Mr. HORN. Are they part of your group?

Mr. WALL. The VA system, at least in Denver, the hospital in Denver is a member of the association. The other outlying VA hospitals are not members. They are appointed to be represented on the hospital preparedness advisory committee, but to date I don't believe we've had a representative attend any of our hospital preparedness advisory committee meetings. With regard to the issue of capacity that you raised earlier, there was a time in Colorado when we had approximately 5 beds per 1,000 population. As a result of the changes in the health care economy, we are down today to about 2 beds per 1,000 population. So the issue of excess capacity is not present any longer, and I think Dr. Miller talked about surge capacity, and we right now as part of the needs assessment process are evaluating how many gurneys are available, can beds be put in hallways, do we have power to be able to do that, and so on and so forth. So I think the issue of capacity with regard to application to a significant event is something of a very significant nature that we need to be very, very concerned about.

For example, most hospitals in Colorado were built in the 1960's and 1970's. We obviously have some that are——

Mr. HORN. Hill-Burton?

Mr. WALL. A lot of Hill-Burton money around the country and Colorado included. Emergency power needs in the 1960's and 1970's were very different than emergency power needs today. I mean, computers weren't even a reality back in those days, and now all of the technology that we have is basically driven by computers. Emergency power was not, in fact, required at that level at that time. It was basically required to be able to continue a surgery in an OR, for hall lighting, perhaps for food preparation, but nowhere near to the extent were the power requirements in place in that time that we have today. So to upgrade capacity in case power had to be self-sustaining is a very significant issue for most hospitals in the eventuality of an event like this.

Work force capacity is a significant issue today. Even if we had all of the equipment that we needed, even if we had all the beds we needed, the question is, where is the work force going to come from to staff those beds if all of a sudden you had a need for 1,000 new patients in a metropolitan area like Denver? I can tell you right now the capacity is not there to address an event of that size.

Dr. Miller was talking about the work that we are doing trying to address this issue on a regional basis where we could take care of 500 patients. That is a lot of additional patients in a system which, quite frankly, from a bed capacity point of view, is there right now.

So there are some very significant needs out there, and that's why I raise the funding issue as important. And clearly the money is needed for the improvement of the infrastructure for public health. There's no question about that. But to put that in context, the first-year grants to public health were about $14 million in the State of Colorado. The first-year HRSA grant for hospitals in the State of Colorado was $1.9 million, a portion of which is utilized for staff required as part of the grant as well as the cost of the needs assessment which was part of the grant as well. So we may net out of that for hospital capability the first year maybe a million and a half dollars. And I can tell you, that doesn't begin to scratch the surface in terms of having hospitals prepared. And the issue of preparedness is ongoing. It's not that we get ready for a year and then we don't have to be concerned about it. This is going to be an issue of ongoing concern well into the future, and I think we need to look at it in a long-term context.

Mr. HORN. I think some of these terrorists will be doing a lot of mischief, and most hospitals have a separate supply of electricity and energy, I think I'm right on that, where they have gasoline and motors, should they ever try to do something like knocking down the towers and all that. And that will probably happen just as it did in New York. But the power goes out and all the rest, and what do we do and what are we doing now to make sure that they can't get to the different batteries that are in many hospitals and motors to generate that energy?

Mr. WALL. I think you have two issues there. Certainly hospitals have emergency capability. I think my point is that the bulk of that emergency capability in hospitals is undersized based upon the power needs of hospitals in the year 2000. And for us to think that we have the capability on those emergency power systems to run all of the equipment that we would need in order to respond to an event like this, I think is not appropriate. It is just not there. Emergency power capability is critically important.

With regard to the issue of security, hospitals throughout this country, and certainly in Colorado, have internal security forces in most cases, although, again, the further outside the metropolitan area one goes, the more problematic that becomes. But there certainly would be ways, I think, of protecting a hospital from outside mischief, for the most part. Those generators are tested periodically, and they are available. But, again, it's a question of the capability of those generation systems that is in question.

Mr. HORN. The first panel, before they went away, we said, if you could see the President for 30 seconds, what's the most important thing you'd tell him?

Mr. WALL. I think there's probably three things. One is the continued need for ongoing work force development. And sometimes that may not be viewed as part of preparedness, but the reality is we can have all the equipment and capacity we need, but if the work force isn't there to provide the care, it's all for naught. So I think work force development is absolutely essential.

Second, the training of that work force is critical. The issue of bioterror agents, chemical agents, and weapons of mass destruction are new things for hospitals to have to deal with. I think—As the lieutenant said earlier, he's been in this 6 years and still feels lost on occasion, and I think the same is true for health care professionals, although that's their daily routine. We are talking about a new and different time and new and different agents, which is going to require extensive training and ongoing training because of the turnover of personnel that occurs as well.

And then last, obviously, the equipment needs that we have. Most hospitals, certainly outside of Metro Denver, do not have individual decontamination capability. There's very little personal protective equipment available. And, obviously, the key thing we've talked about today is the communications system. And in a State like Colorado, with our varied geography, communication in non-emergency circumstances is a tremendous challenge, let alone during an emergency where it's necessary for all first-line responders and enforcement agencies and incident command centers to all be able to talk with one another with regard to the management of that incident. And I can tell you that is a tremendous problem in the State of Colorado.

Mr. HORN. What do you feel are the human, in terms of personnel, type of issues for the hospitals? Is it the nurse shortage?

Mr. WALL. Nursing is clearly an issue, but in Colorado we also have shortages of radiologic technologists, laboratory personnel. Labs have been mentioned a number of times today, and laboratory personnel are in short supply. And another key profession is that of pharmacists. We have a significant shortage of pharmacists throughout the entire country. And if you begin to think about the distribution of pharmaceutical supplies during an event of this nature, they're going to be a very key profession in our ability to respond.

Mr. HORN. What about the various scholarships we have had? Maybe they aren't enough. We've had nursing scholarships from the Federal Government. We've had the GI bill generally after the Second World War. And what do you think the government should do about that? Now we are talking about bringing people from the Philippines and all that, and they're already here. I mean, they've been here for probably two decades at least. How do you think you're going to solve this problem of getting them educated?

Mr. WALL. Well, I think, again, Mr. Chairman, it goes back to the issue of adequate funding for the training of the health professions. And over the years, I think if you look at the expenditure for the education of health professionals, it's progressively become less and less and less. And I know Congress right now is looking at a Nursing Reinvestment Act. I think that's a critical issue, but not only for nursing, but for the other health professions as well. We in health care have our own work to do to make health care an attractive profession for individuals as well. Quite frankly, the ability to move into the computer sciences where you work 5 days a week, 8 hours a day, make significant 6-figure incomes, compared to nursing, where you're working 7 days a week or you're at least available 7 days a week, 24 hours a day, 365 days a year, that's not very enticing to young people. And I think we need to do some things about that from the point of health care as well. But we clearly need to reemphasize the importance of the health care professions at the national level.

Mr. HORN. Thank you. Mr. Sullivan, if you were in the President's oval office for 30 seconds, what would you tell him?

Mr. SULLIVAN. I think I'd just sum up my earlier remarks. Additional resources, flexibility in their use, coordination, and the interoperability of communications equipment, and that would be both the personnel and the technical interoperability. We have a tendency, I think, to focus on solving our communication needs by buying new radio systems. It's been my experience that a lot of the communication doesn't take place because of cultural differences or blinders, if you will, of the agencies involved. And that's something we need to address.

Mr. HORN. So you've really got to work with the culture of bureaucracy, bureaucracy by bureaucracy?

Mr. SULLIVAN. Essentially. And the tendency, particularly on the first responders' level, to focus in on your particular discipline and what you've been trained to do and not step back and realize that it's a broader issue, and you need to communicate with law enforcement, fire, EMS, public health, and those types of things.

Mr. HORN. And with this new department that we have, a lot of mergers and a lot of corporate mergers have occurred. And when those come together, there's often also a problem—just like Customs, 200 years of real help in this country. And same thing with the Coast Guard. You have a problem, and corporations have looked at that. And you've got to be very fair to those you merge with so that the one group does not have all the positions. They've got to meld them with the rest of the group and have them all working together.

Mr. SULLIVAN. Yes, sir. We need to—It's just a simple process—I wish it were a simple process—of breaking down barriers. There's been a lot of talk today about intelligence sharing. That needs to happen. We need to break down a lot of the barriers. That isn't just a Federal and State problem; it's a local problem. Law enforcement—and I say that even though the two gentlemen to my left are armed——

Mr. HOFFNER. I don't have any bullets.

Mr. SULLIVAN [continuing]. Has been reluctant to share outside of the law enforcement community. And it's one thing for the FBI to send information to State law enforcement and local law enforcement, but if that information doesn't get distributed out to public health and EMS and fire and emergency management agencies, it really doesn't serve its purpose.

Mr. HORN. Thank you. What would you tell the President, Lieutenant Hoffner?

Mr. HOFFNER. Well, if I had 30 seconds, I'd say it's really important that we make sure that the Federal Government keeps doing what they're doing, but we need not forget who's going to be the first responders to that incident, and who's going to be taking care of that incident and those people and those victims and those communities for the first 2 or 3 days. And that's our law enforcement and fire departments, and we need to make sure that they have the training and the personal protection equipment to make sure that they can survive.

Mr. HORN. How about you, what's your version, Lieutenant?

Mr. WICKS. Mr. Chairman, I think one of the things I would ask the president to do is assist the civilian population, if you will, with training for these critical incident commanders to include community leaders, some of the politicians, you know, kind of in this same environment, so the community leaders understand the issues that we as critical incident commanders have to deal with, and we would all kind of get the global perspective more than that myopic perspective of this is what I do and there's no cause and effect out there, because there is. What I do will have a cause-and-effect and a ripple effect on a lot of other people and agencies.

Mr. HORN. Mr. Posner, 30 seconds.

Mr. POSNER. Well, I would say the well-intentioned efforts to respond to crises in the past often lead to phenomena that's been called ready, fire, aim, and we ought to be better prepared this time to put management more up front in terms of thinking clearly, what are the tools we are going to use, how are we going to design them, and how we going to deploy managers at the Federal, State, and local level to really think through how these things are going to be implemented.

And in that regard, we had a vehicle where these kinds of discussions took place in Washington—the ACIR, the Advisory Commission on Intergovernmental Relations. It's been out of business for a while. And these field hearings are very useful in bringing up some issues we need to be thinking more systematically about in Washington, and how can we create a forum to have these kind of discussions back there.

Mr. HORN. Well, we thank you all. And I think we've learned quite a bit. And it will be a good report with the help of our fine reporter.

I want to put on the record thanks from the staff, and that includes Mr. Russell George, who is now leaving the subcommittee, but he's done a wonderful job over the last 5 or 6 years. And he is now the inspector general for an agency that—he was given a nomination and now confirmed by the Senate. And so the lady on the left here, and your right, is the acting staff director, Bonnie Heald. We also have the chief of staff in my office, Dave Bartel is back there. And Chris Barkley is assistant to the subcommittee. And Bonnie Heald and Chris Barkley and Dave are all on this particular hearing, and we thank them a lot because it meant tough hours, 2 and 3 in the morning working, this kind of thing.

Michael Sazonov is back in Washington, staff assistant. And here in Denver we have Dan Kopelman and Adam Roth from Representative Tancredo's staff. And then Anne Roelofs is the Jefferson County facilities person, and she got us this wonderful auditorium here. She hasn't towed our cars away yet, we think, and she's a real worker who knows how to do things. And not least but the best is here, the court reporter, Stacy Armstrong. So thank you very much. And with that, we are adjourned.

[Whereupon, at 12:47 p.m., the subcommittee was adjourned.]

○